Transmission Systems II

A textbook covering the Level II syllabus of the Technician Education Council

D C Green
M Tech, C Eng, MIERE
Senior Lecturer in Telecommunication Engineering
Willesden College of Technology

Pitman

Pitman Publishing Limited
39 Parker Street, London WC2B 5PB

Associated Companies
Copp Clark Ltd, Toronto
Fearon Publishers Inc, Belmont, California
Pitman Publishing New Zealand Ltd, Wellington
Pitman Publishing Pty Ltd, Melbourne

© D C Green 1978

First published in Great Britain 1978

All rights reserved. No part of this publication may be reproduced,
stored in a retrieval system, or transmitted, in any form or by any
means, electronic, mechanical, photocopying, recording and/or
otherwise without the prior written permission of the publishers.
This book may not be lent, resold, hired out or otherwise disposed
of by way of trade in any form of binding or cover other than that
in which it is published, without the prior consent of the publishers.
This book is sold subject to the Standard Conditions of Sale
of Net Books and may not be resold in the UK below the net price.

Text set in 10/12 Linotron Times
printed by photolithography and bound in Great Britain
at The Pitman Press, Bath

ISBN 0 273 01124 3

Contents

1 Frequency, Wavelength and Velocity · 1
The voice and speech · 1
The transmission of sound over a simple telephone circuit · 4
Television signals · 6
Telegraph signals · 8
Range of frequencies in communication · 11
Phase velocity, frequency and wavelength · 12
Exercises · 15

2 Modulation · 17
Frequency-division multiplex · 18
Time-division multiplex · 20
Types of modulation · 22
Amplitude modulation · 25
Signal-to-noise ratio · 34
Single-sideband operation · 35
Frequency modulation · 37
Data transmission · 41
Exercises · 45

3 Carrier Frequencies, Bandwidths and Maximum Power Transfer · 48
Line communication systems · 48
Radiocommunication systems · 53
Maximum power transfer · 57
Exercises · 60

4 Filters · 62
Inductor-capacitor filters · 62
Crystal filters · 68
Filters in parallel · 70
Active filters · 71
Exercises · 72

5 The Decibel · 73
The decibel · 75
The neper · 81
Measurement of decibels · 82
Exercises · 85

6 Telephone Lines and Cables · 88
Basic transmission line theory · 88
Group velocity · 94
Construction of telephone cables · 98
External cables · 98
Internal cables · 105
Loading of cables · 106
Use of transmission lines as radio station feeders · 108
The effect of cables on analogue and digital signals · 109
Exercises · 110

7 Two-wire and Four-wire Circuits · 113
Local lines · 114
Junction and trunk circuits · 115
Two-wire circuits · 115
The interconnection of junction and trunk circuits · 119
Ship radio-telephones · 122
Exercises · 122

8 Frequency-division and Time-division Multiplex Systems · 124
Balanced modulators · 124
Frequency-division multiplex systems · 127
Time-division multiplex systems · 132
Relative merits of f.d.m. and t.d.m. · 135

Numerical Answers to Exercises · 138
Learning Objectives · 139
Index · 140

Preface

This book provides an introductory course on the basic principles of telecommunication transmission systems and is intended for the telecommunication technician. Transmission systems are an integral part of all national and international telecommunication networks and an understanding of their operation is necessary for all technicians, whether their particular interests lie in the field of line or radio communication or telephony.

The Technician Education Council (TEC) scheme for the education of telecommunication technicians consists of a number of standard units which are studied over a period of three years. This book provides complete coverage of the level II unit Transmission Systems II. Chapter 1 discusses the range of frequencies produced by the voice, by musical instruments, and by other sources of information. Chapter 2 then introduces the reader to the need for, and the principles of, modulation before Chapter 3 considers the choice of carrier frequency and bandwidth for practical systems. Chapters 4 and 5 then consider, respectively, the basic concepts of electric filters and the meaning and use of the decibel. The elementary theory of transmission lines and the construction of cables are the subject matter of Chapter 6. The final two chapters in the book are concerned with the ways in which point-to-point line links can be operated and in so doing introduce the reader to the principles of multi-channel telephony systems.

The book therefore provides a text which gives a comprehensive introduction to the basic principles of transmission systems and it should be eminently suitable for any non-advanced course in telecommunication engineering.

The contents of the unit Transmission Systems II have been written by the TEC in the form of learning objectives and these are given at the end of the book. Acknowledgement is made to the TEC for their permission to use the content of the unit. The Council reserve the right to amend the content of its unit at any time.

Many worked examples are provided in the text to illustrate the principles which have been discussed and each chapter concludes with a number of exercises. Many of the examples have been taken from past City and Guilds examination papers and grateful acknowledgement of their permission to do so is made. The answers given to the numerical problems are the sole responsibility of the author and are not necessarily endorsed by the Institute.

D. C. G.

1 Frequency, Wavelength, and Velocity

In telecommunication engineering, media are provided for the transmission of intelligence from one point to another; for example, telephone conversations pass through telephone cables and radio programmes are broadcast through the atmosphere. It is necessary to ensure that sufficient information is available at the receiving end of a system to allow the intelligence to be understood and appreciated by the person receiving it. The requirements demanded of the transmission media depend upon the type of intelligence to be transmitted, but for the transmission of speech, music and television the main requirement is that sufficient frequencies are retained in the transmitted waveform to permit the received sound and picture to be understandable and, in the case of music and television, to be enjoyable also. It is therefore necessary to have an appreciation of the range of frequencies produced by the human voice, by musical instruments and by television systems, and the frequency range over which the human ear is capable of responding.

The Voice and Speech

A current of air expelled by the lungs passes through a narrow slit between the vocal cords in the larynx and causes them to vibrate. This vibration is then communicated to the air via various cavities in the mouth, throat and nose. The shape and size of the nose cavities are more or less fixed, but the mouth and throat cavities can have their shapes and dimensions considerably changed by the action of the tongue, lips, teeth and the throat muscles. The frequency at which a particular cavity allows the air to vibrate most freely depends upon the shape and dimensions of the cavity and these can readily be adjusted by the movement of the lips, tongue and teeth. The pitch of the spoken sound depends upon both the length and tension of the vocal cords and the width of the slit between them. The length of the vocal cords varies from person to person; for example a woman will have shorter vocal cords than a man, while the tension and distance apart of the vocal cords is under the control of muscles.

When a person speaks, his vocal cords vibrate and the resulting sounds, which are rich in harmonics, but of almost constant pitch, are carried to the cavities in the mouth, throat and nose. Here the sounds are given some of the characteristics of the desired speech by the emphasizing of some of the harmonics contained in the sound waveforms and the suppression of others. Sounds produced in this way are the vowels, a, e, i, o and u, and contain a relatively large amount of sound energy. Consonants are made with the lips, tongue and teeth and contain much smaller amounts of energy and often include some relatively high frequencies.

The sounds produced in speech contain frequencies which lie within the frequency band 100–10 000 Hz. The pitch of the voice is determined by the fundamental frequency of the vocal cords and is about 200–1000 Hz for women and about 100–500 Hz for men.

The power content of speech is small, a good average being of the order of 10–20 microwatt. However, this power is not evenly distributed over the speech frequency range, most of the power being contained at frequencies in the region of 500 Hz for men and 800 Hz for women.

Music

The notes produced by musical instruments occupy a much larger frequency band than that occupied by speech. Some instruments, such as the organ and the drum, have a fundamental frequency of 50 Hz or less while many other instruments, for example the violin and the clarinet, can produce notes having a harmonic content in excess of 15 000 Hz. The power content of music can be quite large. A sizeable orchestra may generate a peak power somewhere in the region of 90–100 watts while a bass drum well thumped may produce a peak power of about 24 watts.

Hearing

When sound waves are incident upon the ear they cause the ear drum to vibrate. Coupled to the ear drum are three small bones which transfer the vibration to a fluid contained within a part of the inner ear known as the cochlea. Inside the cochlea are a number of hair cells and the nerve fibres of these are activated by vibration of the fluid. Activation of these nerve

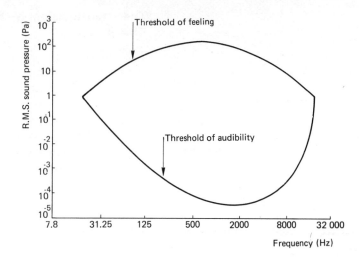

Fig. 1.1 The thresholds of audibility and feeling

fibres causes them to send signals, in the form of minute electric currents, to the brain where they are interpreted as sound.

The ear can only hear sounds whose intensity lies within certain limits; if a sound is too quiet it is not heard and, conversely, if a sound is too loud it is felt rather than heard and causes discomfort or even pain. The minimum sound intensity, measured in Pascals ($1\,\text{Pa} = 1\,\text{N/m}^2$), that can be detected by the ear is known as the "threshold of hearing or audibility" and the sound intensity that just produces a feeling of discomfort is known as the "threshold of feeling". The ear is not, however, equally sensitive at all frequencies, as shown in Fig. 1.1. In this diagram curves have been plotted showing how the thresholds of audibility and feeling vary with frequency for an average person.

It can be seen that the frequency range over which the average human ear is capable of responding is approximately 30–16 500 Hz, but this range varies considerably with the individual. The ear is most sensitive in the region of 1000 to 2000 Hz and becomes rapidly less sensitive as the upper and lower limits of audibility are approached. The limits of audibility are clearly determined not only by the frequency of the sound but also by its intensity. See, for example, the increase in the audible frequency range when the sound intensity is increased from, say, $1 \times 10^{-3}\,\text{Pa}$ to $1 \times 10^{-2}\,\text{Pa}$. At the upper and lower limits of audibility the thresholds of audibility and feeling coincide and it becomes difficult for an observer to distinguish between hearing and feeling a sound.

4 FREQUENCY, WAVELENGTH, AND VELOCITY

The Transmission of Sound over a Simple Telephone Circuit

The intensity of a sound wave rapidly diminishes as it travels away from the source producing it, and if conversation over a long distance is desired a telephone circuit becomes necessary.

The arrangement of a simple, UNIDIRECTIONAL telephone circuit is shown in Fig. 1.2.

It is sufficient to consider the microphone as a device whose electrical resistance varies in accordance with the waveform of the sound incident upon it, and to consider the receiver as a device which vibrates when a varying current is passed through it and in so doing produces sound waves having the same waveform as the current.

When no sound is incident upon the microphone, the resistance of the microphone is constant and a steady current flows into the line from the battery. When a person speaks into the microphone, its resistance varies in the same way as the speech waveform and so does the current flowing to line. For example, during one half-cycle of the speech waveform the resistance of the microphone is decreased and so the line current is increased, while during the next half-cycle the microphone resistance is increased and the line current decreased. Thus the line current is continuously varying about its steady value (Fig. 1.3). The varying line current passes through the receiver at the distant end of the line and causes the receiver to vibrate and reproduce the original speech.

For a TWO-WAY conversation this simple arrangement would have to be duplicated, the second circuit having the positions of the microphone and battery and the receiver reversed. Such an arrangement would be uneconomic since it would require two pairs in a telephone cable, and telephone cables are very expensive. A further disadvantage of the simple circuit is that as the length of line is increased, the variation in the resistance of the microphone becomes an increasingly smaller fraction of the line resistance. This means that the magnitude of the changes in line current decreases with increase in line length until the current changes can no longer operate the receiver.

A telephone circuit which overcomes these disadvantages is shown in Fig. 1.4. In this circuit, each microphone is connected to the line via a transformer but the two receivers are directly connected to line. When a person speaks into either of the two microphones, the resulting changes in microphone resistance cause a relatively large varying current to flow in the local microphone circuit. In passing through the transformer primary winding, this varying current induces an e.m.f. in the secondary winding and the induced e.m.f. drives a varying

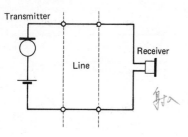

Fig. 1.2 A simple unidirectional speech circuit

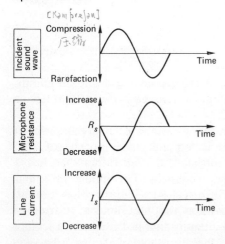

Fig. 1.3 The relationship between the sound incident on a microphone, the microphone resistance, and the current flowing to line. R_s = microphone resistance when no sound is incident upon it. I_s = steady current to line when no sound is incident on the microphone

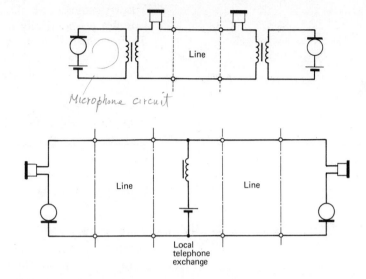

Fig. 1.4 A simple two-way speech circuit

Microphone circuit

Fig. 1.5 The central battery system

current, having the same waveform, to line. At the distant end of the line the varying current flows in the receiver and causes it to vibrate and so produce sound waves which are similar to the original speech.

The circuit given in Fig. 1.4 has the disadvantage of requiring a separate battery at each telephone. The vast majority of telephone circuits are connected to a local telephone exchange and these circuits are all operated from a CENTRAL BATTERY. The basic principle of the central battery system is illustrated by Fig. 1.5. A large capacity secondary cell battery is installed at the local telephone exchange which supplies current to the lines when the telephone handsets are lifted from their rests. When either of the microphones is spoken into, a speech-frequency current is superimposed upon the battery current and is transmitted, via the telephone exchange, to the other telephone. The inductor is connected in series with the exchange battery to prevent the speech-frequency currents entering the battery. In practice, a more complex telephone circuit is needed because the circuit as shown would suffer from the speech-frequency currents generated by the microphone flowing in the associated receiver. The sound picked up by the microphone would be clearly audible in the receiver to give excessive "sidetone".

The telephone network of Great Britain is divided into local lines, junctions and trunks. *Local lines* consist of pairs of wires that connect the individual telephone subscribers to their local telephone exchange; *junctions* are two-wire circuits that may or may not be amplified and connect nearby telephone exchanges together; and *trunks* are amplified four-wire circuits that connect distant exchanges together.

Long-distance telephone lines are extremely expensive and it is not economically possible to connect every exchange in the network to every other exchange; direct trunks are only provided between two exchanges when justified by the traffic carried. Trunk circuits are always amplified and in the majority of cases are routed over one or more multi-channel carrier telephony systems. International circuits may be routed over submarine cable, microwave radio or satellite multi-channel systems, or sometimes over high-frequency (3–30 MHz) radio links.

Television Signals

A television picture is divided vertically into 625 lines, each of which is effectively subdivided into a number of sections. A television picture is therefore divided into a number of elemental areas. If the elemental areas are small enough each will have a constant brightness; for monochrome television, the brightest areas are white, the darkest black, and all other areas are various shades of grey. The action of the television camera is to convert the brightness of each area into a voltage whose amplitude is proportional to the brightness.

The camera is unable to transmit all the voltages simultaneously, and so the system is arranged to transmit them sequentially. The picture is SCANNED by the camera in a series of lines as in Fig. 1.6. The scan starts at the top left-hand corner of the picture, point A, and travels along the first line to point B. As each elemental area is scanned a proportional instantaneous voltage is transmitted and a voltage waveform representing the variation of brightness along the first line is produced. At the end of this line the scan flies rapidly back to point C before travelling along the second line to point D, and so on until the entire picture has been scanned. When the end of the last line has been reached the scan is moved back to point A ready to commence scanning the next field.

At the television receiver the picture is built up from a number of lines that are traced out in sequence by a spot of light travelling on the screen of the cathode-ray tube. The spot is produced by an electron beam incident on the inner face of the screen, and to obtain the various shades of grey, and black and white, demanded by a particular picture the brightness of the spot is modulated as it travels, by the picture signal. For the picture to be reproduced correctly it is essential for the two scans (camera and receiver) to be in synchronism. The picture signal produced by the television camera is therefore accompanied by SYNCHRONIZING PULSES (Fig. 1.7). The persistence of vision of the human eye enables the viewer to see a complete picture and not a moving spot of light. The fields

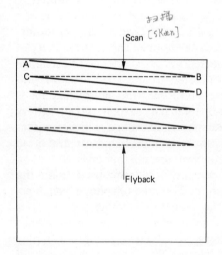

Fig. 1.6 Sequential scanning of a television picture

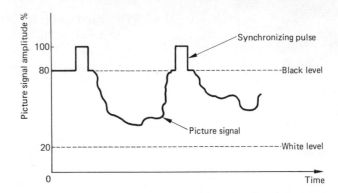

Fig. 1.7 Television picture waveform

must be repeated at a frequency high enough for the slight changes in successive fields to give the impression of movement without apparent flicker. The necessary field frequency can be reduced by 50% if *interlaced scanning* is used; here the odd lines, 1, 3, 5, 7, etc., are scanned first and the even lines afterwards.

The bandwidth required for a television picture signal depends upon a number of factors, such as the number of lines that make up the picture, the number of fields transmitted per second and the durations of the synchronizing pulses. If the synchronizing pulses are neglected an approximate expression for the necessary bandwidth can be obtained.

Let the height of the picture be H; the width of the picture, W; the aspect ratio (picture-width/picture-height) of the picture, x; and the number of lines comprising the pictures, L.

Then the height of each elemental area of the picture is H/L. For the pictures to have equal vertical and horizontal definitions the elemental areas are square; hence each area has a width of H/L.

The picture width W is equal to the aspect ratio times the height of the picture, i.e. $W = xH$, and the number of elemental areas in one line is

$$\frac{\text{Picture width}}{\text{Width of one elemental area}} = \frac{xH}{H/L} = xL$$

The instantaneous voltages produced in the television camera as the picture is scanned are proportional to the brightness of the elemental areas. If all the elements in a line have the same brightness the line voltage produced is of constant amplitude and therefore of zero frequency. The line voltage of maximum frequency is produced when, as in a facsimile telegraphy system, a chessboard pattern is transmitted. The maximum frequency f_{max} is then equal to half the number of elements in a line divided by the time t required to scan one line; i.e.

$$f_{max} = \frac{xL}{2t} \quad \text{hertz} \tag{1.1}$$

The bandwidth required for a television system is thus from zero to $xL/2t$ hertz. Therefore

$$\text{Bandwidth required} = \frac{xL}{2t} \quad \text{hertz} \qquad (1.2)$$

EXAMPLE 1.1

The British 625 line television system has an aspect ratio of 4/3 and a field frequency of 25 fields per second. Calculate the bandwidth necessary to transmit this system, assuming that equal vertical and horizontal definition is required.

Solution
The time taken to scan one line is

$$t = \frac{1}{\text{Field frequency} \times \text{Number of lines}} = \frac{1}{25 \times 625} \text{ sec}$$

Substituting in equation (1.2),

$$\begin{aligned}
\text{Bandwidth required} &= \frac{\frac{4}{3} \times 625}{\frac{2 \times 1}{25 \times 625}} \\
&= \frac{4 \times 625 \times 25 \times 625}{2 \times 3} \\
&= 6.51 \text{ MHz} \quad (Ans.)
\end{aligned}$$

If the time durations of the synchronizing pulses are taken into account it is found that bandwidth actually required is 7.37 MHz. In practice, the normal bandwidth is 5.5 MHz, resulting in some loss of horizontal definition.

A colour television picture signal consists of a brightness (*luminance*) component, which corresponds to the monochrome signal previously described, plus colour (*chrominance*) information which is transmitted as the amplitude-modulation sidebands of two colour *subcarriers* which are of the same frequency (approximately 4.434 MHz) but are 90° apart. No extra bandwidth is needed to accommodate the colour information.

Telegraph Signals

Telegraphy is the passing of messages by means of a signalling code. Several different codes are in use for different telegraphy systems, but the two most common are the *Morse code* and the *Murray code*.

In the MORSE CODE characters are represented by a combination of *dot* signals and *dash* signals; the difference between a dash and a dot is one of time duration only, a dash having a period three times that of a dot. Spacings between

FREQUENCY, WAVELENGTH, AND VELOCITY 9

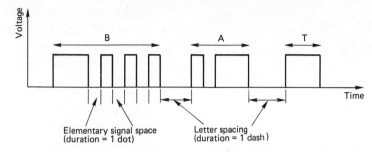

Fig. 1.8 BAT in Morse code

elementary signals, between letters and between words are also distinguished from one another by different time durations. As an example, Fig. 1.8 shows the word BAT in Morse code. The Morse code is not convenient for use with automatic-printing receiving equipment because the number of signal elements needed to indicate a character is not the same for all elements, and the signal elements themselves are of different lengths.

In the MURRAY CODE all characters have exactly the same number of signal elements and the signal elements are of constant length. Each character is represented by a combination of five signal elements that may be either a *mark* or a *space*. In Great Britain a mark is represented by a negative potential or the presence of a tone, and a space is represented by a positive potential or the absence of a tone. Fig. 1.9 shows the letters B, T, R and Y in Murray code. The Murray code is used for all teleprinter systems.

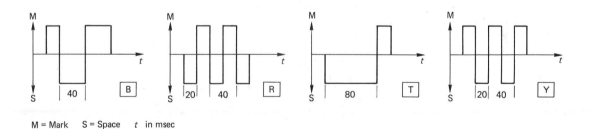

M = Mark   S = Space   *t* in msec

Fig. 1.9 B, T, R and Y in Murray code

Telegraph speed is measured in terms of a unit known as the *baud*. The baud speed of a telegraph signal is the reciprocal of the time duration of the shortest signal element employed. In the Morse code the dot is the shortest signal element; in the Murray code the elements are all of the same length. The bandwidth required for the transmission of a signal in Morse code depends upon the number of words transmitted per minute and has not been standardized, but generally lies in the range 100–1000 Hz. Teleprinters are normally operated at a telegraph speed of 50 bauds, and this means that the time duration of a mark, or of a space, is 1/50 second, or 20 ms. The bandwidth required to transmit a teleprinter signal depends upon the characters sent, but the maximum bandwidth is demanded when alternate marks and spaces are transmitted, i.e. letters R and Y. The periodic time of one cycle of the waveforms for R and Y is 40 ms, and so the fundamental frequency of the waveform is 1000/40, or 25 Hz.

Picture, or Facsimile, Telegraphy

Facsimile telegraphy is the transmission and reception of still pictures, diagrams, etc. The basic principle of a facsimile telegraphy system is shown in Fig. 1.10.

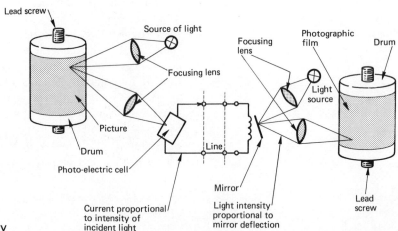

Fig. 1.10 Facsimile telegraphy

The picture to be transmitted is mounted on a circular drum that is free to move along a lead-screw as it revolves about it. A small spot on the picture is illuminated by the focused light from a light source and, as the drum revolves, all parts of the picture are successively scanned by the spot of light. The light reflected from the illuminated spot is focused on to a photo-electric cell that gives a voltage output proportional to the intensity of the light incident upon it. As the picture is scanned, the light reflected from the illuminated spot fluctuates according to the nature of the picture and so the current delivered by the cell to the line varies also.

At the receiving end of the system a photographic film is scanned in similar fashion by a light beam whose intensity is proportional to the magnitude of the received current. Light from a light source is focused on the surface of a mirror whose position is a function of the magnitude of the photo-cell current passing through an inductive winding. The intensity of

FREQUENCY, WAVELENGTH, AND VELOCITY

the light reflected from the mirror, and focused on the photographic film, depends upon the angle of the mirror. If the system is correctly adjusted each part of the film is illuminated as it is scanned just sufficiently for the original picture to be reproduced.

The bandwidth required for a facsimile telegraphy system depends upon the speed with which pictures are transmitted and the nature of the pictures, but usually a bandwidth between 0–500 Hz and 0–1000 Hz is required for direct transmission.

Facsimile telegraphy is employed for the transmission of Press photographs but it is not economical, compared with teleprinters, for the transmission of written messages.

Range of Frequencies in Communication

It has been assumed so far that all the frequencies present in a speech or music waveform are converted into electrical signals, transmitted over the communication system, and then reproduced as sound at the distant end. In practice, this is rarely the case for two reasons. Firstly, for economic reasons the devices used in circuits that carry speech and music signals have a limited bandwidth; secondly, particularly for the longer-distance routes, a number of circuits are often transmitted over a single telecommunication system and this practice provides a further limitation of bandwidth. It is thus desirable to have some idea of the effect on the ear when it is responding to a sound waveform, the frequency components of which have amplitude relationships differing from those existing in the original sound.

By international agreement the audio frequency band for a "commercial quality" speech circuit routed over a multi-channel system is restricted to 300–3400 Hz; while for a circuit operated over a high-frequency radio link the bandwidth is only 25–3000 Hz. This means that both the lower and upper frequencies contained in the average speech waveform are not transmitted. To the ear, the pitch of a complex, repetitive sound waveform is the pitch corresponding to the frequency difference between the harmonics contained in the waveform, i.e. the pitch is that of the fundamental frequency. Hence, even though the fundamental frequency itself may have been suppressed, the pitch of the sound heard by the listener is the same as the pitch of the original sound. However, much of the power contained in the original sound is lost. Suppression of all frequencies above 3400 Hz reduces the quality of the sound but does not affect its intelligibility. Since the function of a telephone system is to transmit intelligible speech, the loss of quality can be tolerated; sufficient quality remains to allow a speaker's voice to be recognized.

12 FREQUENCY, WAVELENGTH, AND VELOCITY

In the transmission of music the enjoyment of the listener must also be considered. The enjoyment of a person listening to music over a communication system may be considerably impaired if too many of the higher harmonics have been suppressed, since the loss of these harmonics could well result in it becoming difficult, if not impossible, to distinguish between the various kinds of musical instruments. For music circuits routed over line communication systems, a wider bandwidth must be allowed than is allocated to commercial speech circuits. A typical bandwidth in practice is 30 to 10 000 Hz, which makes excellent quality reproduction possible. Land-line music circuits of this kind are used to connect together two BBC studios or a studio and a transmitting station. When music is broadcast in the long and medium wavebands a bandwidth as wide as 30–10 000 Hz cannot be achieved because the wavebands are shared by so many different broadcasting stations.

By international agreement medium waveband broadcasting stations in Europe are spaced approximately 9000 Hz apart in the frequency spectrum, and this means that to make it possible for any particular station to be selected by a radio receiver, without undue interference from adjacent (in frequency) stations, the output sound bandwidth of the receiver cannot be much greater than 4500 Hz. Thus the effective bandwidth of a medium wave broadcast transmission, be it music or speech, is of the order of 50–4500 Hz. Sound broadcast stations in the high-frequency band are allowed an r.f. bandwidth of 10 kHz. This provides an audio bandwidth of 50–5000 Hz. For various reasons the same selectivity is not demanded of very high frequency (v.h.f.) sound broadcast receivers or of ultra high frequency (u.h.f.) television receivers (625 line), and consequently the bandwidths handled by these receivers are somewhat greater.

Frequency-modulated sound broadcast transmissions in the 88–108 MHz band provide audio signals up to 15 kHz which allows reasonably good quality reception of musical programmes with a receiver of adequate performance. The sound signals provided with u.h.f. television programmes also have a 15 kHz audio bandwidth.

The video bandwidth allocated in practice to a u.h.f. television signal is not as high as calculated earlier but, for economic reasons, is limited to 5.5 MHz. This practice leads to some loss of the horizontal definition of the received picture.

Phase Velocity, Frequency, Wavelength

The propagation of a voltage or current wave along a telephone line, or of a radio wave through the atmosphere, is not an instantaneous process but occupies a definite interval of time. There is thus a time lag between the application of a voltage at one end of a line and the detection of a voltage change at a distant point along the line. This means that there will be a PHASE DIFFERENCE between the a.c. voltages existing at two points along a line at a given instant, since the instantaneous voltage at each point is continuously changing. For example, suppose that for a given line the voltage at a point x miles from the sending end lags behind the sending-end voltage by 90°. Then at a distance of $2x$ miles from the sending end the voltage will be lagging by $2 \times 90°$ or 180° on the sending-end voltage. The voltage $4x$ miles from the sending end will be in phase with the sending-end voltage, i.e. a complete cycle of instantaneous values has been completed. The distance $4x$ is known as *one wavelength*, symbol λ (see Fig. 1.11). The wavelength λ is the distance, in metres, between two similar points, in the propagating waveform, e.g. the distance between two successive positive peaks as shown.

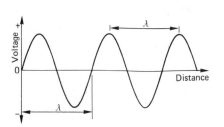

Fig. 1.11 Wavelength of a sinusoidal signal

Suppose a sinusoidal voltage of frequency f is applied to the input terminals of a transmission line. The periodic time of this waveform is $T = 1/f$ seconds. In a time T seconds the voltage wave will travel along the line a distance equal to T times the velocity v with which it is propagated. During this time the voltage at the sending end of the line will have described a complete cycle of variation with time. While this happens the wave travels a distance that is the wavelength λ of the signal. Therefore

$$\lambda = vT = \frac{v}{f}$$

or

$$v = \lambda f \tag{1.3}$$

A radio wave travelling through the atmosphere has its velocity of propagation equal to the velocity of light c. In this book c will be taken as 3×10^8 m/s. The velocity of propagation of a voltage, or current, wave along a transmission line is less than 3×10^8 m/s, the amount of reduction depending upon the inductance and capacitance of the line.

EXAMPLE 1.2

A 30 MHz radio wave is propagated through the atmosphere. Calculate its wavelength.

Solution
From equation (1.3)

$$\lambda = \frac{3 \times 10^8}{30 \times 10^6} = 10 \text{ m} \quad (Ans.)$$

14 FREQUENCY, WAVELENGTH, AND VELOCITY

EXAMPLE 1.3

The BBC broadcasting station Radio 3 is broadcast on a wavelength of 464 m. What is its frequency?

Solution

$$f = \frac{3 \times 10^8}{464} = 647 \text{ kHz} \quad (Ans.)$$

EXAMPLE 1.4

A 5 MHz radio transmitter is connected to an aerial by a length of transmission line. If the wavelength of a signal on the transmission line is 0.8 times the wavelength of a signal in the atmosphere, calculate the velocity with which a signal is propagated along the cable.

Solution
For the wave in the atmosphere,

$$\lambda_{air} = \frac{c}{f} = \frac{3 \times 10^8}{5 \times 10^6} = 60 \text{ m}$$

Hence, the wavelength in the cable is

$$\lambda_{cab} = 0.8 \times 60 = 48 \text{ m}$$

and therefore the velocity of propagation in the transmission line is

$$v_p = \lambda_{cab} f = 48 \times 5 \times 10^6 = 2.40 \times 10^8 \text{ m/s} \quad (Ans.)$$

EXAMPLE 1.5

If the maximum frequency tolerance for fixed service stations operating in the frequency band 4.0 to 27.5 MHz is ±30 parts in 10^6 (±0.003%) and for television stations operating in the frequency band 40.0 to 70.0 MHz is ±10 parts in 10^6 (±0.001%), to what frequency variations in hertz do these tolerances correspond for

(a) a fixed service station on 12.3 metres,
(b) a television station on 5.5 metres?

At what wavelength does a fixed-service station, having a variation of ±300 Hz, become outside the permitted tolerance? (C&G)

Solution
$f = v/\lambda$, where f = frequency in hertz, v = velocity in metres/sec, and λ = wavelength in metres.
 (a) $\lambda = 12.3$ m:

$$f = \frac{3 \times 10^8}{12.3} = 24.4 \text{ MHz}$$

The maximum frequency tolerance in this band is ±30 parts in 10^6.

$$\text{Frequency tolerance} = \pm \frac{30}{10^6} \times 24.4 \times 10^6 = \pm 732 \text{ Hz} \quad (Ans.)$$

(b) $\lambda = 5.5$ m:

$$f = \frac{3 \times 10^8}{5.5} = 54.55 \text{ MHz}$$

The maximum frequency tolerance in this band is ± 10 parts in 10^6.

Frequency tolerance $= \pm \dfrac{10}{10^6} \times 54.55 \times 10^6 = 545.5$ Hz (*Ans.*)

Last part:
Frequency variation allowed = Frequency × Frequency tolerance
Therefore

$$\pm 300 = \text{Frequency} \times \pm \frac{30}{10^6}$$
$$\text{Frequency} = \frac{300 \times 10^6}{30} = 10 \text{ MHz}$$

Therefore

$$\lambda = \frac{3 \times 10^8}{10 \times 10^6} = 30 \text{ m} \quad (Ans.)$$

Exercises

1.1. Speech is to be transmitted from one point to another by electrical means and without the intervention of an exchange. Draw a circuit diagram of a suitable arrangement. State the functions and the general principle of operation of each of the devices used in the arrangement you adopt.

State, in general terms, the effect of any line resistance on the performance of your arrangement. (C&G)

1.2. Give a brief description of the means whereby sound waves are converted into electric currents suitable for transmission along a pair of wires and reconverted at the receiving end into sound waves. (C&G)

1.3. Explain briefly the relationship between the frequency and wavelength of a radio wave. Broadcasting stations work at a constant frequency separation of about 10 kHz. What difference in wavelength separates stations when their frequencies are 1000 kHz and 1010 kHz? (C&G)

1.4. State the relationship between the velocity, frequency and wavelength of radio waves, indicating the units employed. What wavelength corresponds to the frequencies: (*a*) 600 kHz, (*b*) 9 MHz and (*c*) 50 MHz?

State a suitable application for each frequency. (C&G)

1.5. In some transmission systems the following frequency ranges are employed: (*a*) for speech 300–3400 Hz, (*b*) for music 50–6000 Hz, (*c*) for television 50 Hz–3 MHz.

Explain why the bandwidths are so limited and discuss the disadvantages which arise.

FREQUENCY, WAVELENGTH, AND VELOCITY

1.6. What average range of frequency is (*a*) produced by the human voice, (*b*) detected by the human ear?
  Describe briefly how vowel sounds and consonants are produced and give the approximate ranges in frequency which occur in them. Thence, or otherwise, discuss the effects on the transmission of speech which would be caused by eliminating all frequencies (i) above 1500 Hz, (ii) below 1500 Hz. (C&G)

1.7. A certain transmission system has a bandwidth of 10 MHz. Explain why the number of television channels that can be transmitted simultaneously over the system is much less than the number of speech circuits. (C&G)

1.8. (*a*) State the frequency bandwidths required to transmit by radio the following: (i) commercial speech, (ii) telegraphy by teleprinter, (iii) 625 line television.
  (*b*) What are the essential differences between the methods of transmission of speech over a broadcast network and through a high-frequency radio system?
  (*c*) What bandwidth is allocated to a speech channel in a line transmission system? (part C&G)

1.9. Explain the following terms with reference to the United Kingdom television broadcasting service: (*a*) raster, (*b*) line frequency, (*c*) field frequency, (*d*) interlacing, (*e*) flyback, (*f*) synchronization, (*g*) definition. (C&G)

1.10. (*a*) A 30 kHz radio signal is propagated through the atmosphere between two points 45 km apart. Express the distance separating the two points in terms of wavelengths.
  (*b*) Plot a graph to show how the wavelength of a signal varies as its frequency is varied from 500 kHz to 1500 kHz in 250 kHz steps.

Short Exercises

1.11. List the appropriate bandwidths of telegraph, speech, music and video (television picture) signals.

1.12. A 30 MHz signal is transmitted through a cable with a velocity 0.86 times the velocity of light. Determine the wavelength of the signal.

1.13. A transmission line has an electrical length of $\lambda/4$ at a particular frequency. How long will the line be if the frequency is (i) doubled and (ii) trebled?

1.14. Determine the wavelengths of the following signals assuming a velocity of propagation equal to that of light: (*a*) 50 Hz mains supply, (*b*) a 100 kHz radio-telephony signal, (*c*) 1 MHz medium-wave radio broadcast station, (*d*) a v.h.f. frequency-modulated sound broadcast station operating on 90 MHz, (*e*) a u.h.f. colour television channel operating at 800 MHz and (*f*) a microwave equipment working at 6 GHz.

1.15. Explain briefly the meanings of the following terms used in conjunction with the transmission and reception of sound waves: (i) threshold of feeling, (ii) Pascal, (iii) wavelength and (iv) pitch.

1.16. What is the range of frequencies produced by a woman's voice? What is the range of frequencies heard by the average person? What is the effect on the received sound if the transmitted speech has its bandwidth limited to 300–3400 Hz?

1.17. If 625 line television systems did not use interlaced scanning the field frequency would have to be doubled to 50 Hz. Determine the video bandwidth which would then be needed.

2 Modulation

The bandwidth required for the transmission of commercial-quality speech is 300–3400 Hz and as long as a physical pair of wires, or two pairs of wires, are provided to connect the two parties to a conversation no problem exists. Telephone cable is, however, very expensive and, together with the associated duct work and manholes, comprises the major part of the cost of linking two points. It is desirable, therefore, to be able to transmit more than one conversation over a given link and thus economize in telephone cable. If a number of telephone conversations were merely transmitted into one end of a line it would not be possible to separate them at the distant end of the line since each conversation would be occupying the same part of the frequency spectrum. How then can many different conversations be transmitted over a single circuit and yet be separable at the distant end? Two main methods exist: in one, known as Frequency-division Multiplex, each conversation is shifted, or translated, to a different part of the frequency spectrum. In the other, known as Time-division Multiplex, the conversations each occupy the same frequency band but are applied in sequence to the common line.

Frequency-division Multiplex

To illustrate the principle of a frequency-division multiplex (f.d.m.) system, consider the simple case where it is required to transmit three telephone channels, of bandwidth 300 to 3400 Hz, over a common line. The first of these channels can be transmitted directly over the common line and it will then occupy the band 300 to 3400 Hz. The second and third channels cannot also be transmitted directly over the line because they would be inseparable from the first channel and from each other. Suppose, therefore, that instead these two channels are each passed into a circuit which frequency translates, or shifts, them to the frequency bands 4300 to 7400 Hz and 8300 to 11 400 Hz respectively, before transmission to line. The three channels can now all be transmitted over the common line but since there is a frequency gap of 900 Hz between them no inter-channel interference will occur. At the receiving end of the line, filters separate the three channels and further circuits restore the second and third channels to their original frequency bands. The bandwidth provided by the common circuit must be 300–11 400 Hz.

The frequency translation of a channel to a position higher in the frequency spectrum is known as MODULATION and the circuit which achieves it as a modulator. The particular part of the frequency spectrum to which the channel is shifted is determined by the frequency of the sinusoidal *carrier wave* which is modulated. Modulation may be defined as the process by which one of the characteristics of a carrier wave is modified in accordance with the characteristics of a modulating signal. The restoration of a channel to its original position in the frequency spectrum is known as demodulation, the device is known as a demodulator.

A block schematic diagram of the equipment required for one direction of transmission in the three-channel f.d.m. system just described is shown in Fig. 2.1. It should be noted that since the equipment is unidirectional, it needs to be duplicated for transmission in both directions. Furthermore, the line may be a telephone cable or may be a v.h.f., u.h.f. or microwave radio link.

Frequency translation is also employed for radio and television broadcasting. It is well known that such broadcasts are transmitted by the broadcasting authority and received in the home by means of aerials, but no type of aerial is able to operate at audio frequencies. It is therefore necessary to shift each programme originally produced in the audio-frequency band to a point higher in the frequency spectrum where aerials can operate with reasonable efficiency. Since there is usually a large number of broadcasting stations within a given geog-

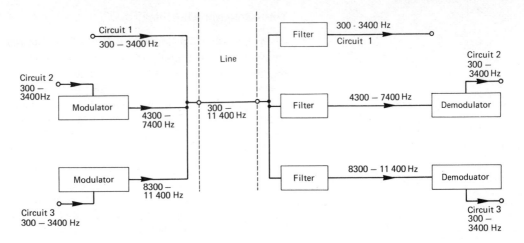

Fig. 2.1 A simple 3-channel f.d.m. system

raphical area, it is necessary to arrange that each station, as far as possible, occupies a different part of the usable frequency spectrum. Hence the programmes radiated by different broadcasting stations are frequency translated to their own, internationally agreed, fixed frequency bands. For example, consider the medium-wave BBC programmes Radios 1, 3 and 4. Radio 1 is broadcast on a frequency of 1214 kHz, Radio 3 on 647 kHz, and Radio 4 on 692, 908, and 1052 kHz.

Time-division Multiplex

With time-division multiplex (t.d.m.) a number of different channels can be transmitted over a common circuit by allocating the common circuit to each channel in turn for a given period of time, i.e. at any particular instant only one channel is connected to the common circuit. The principle of a t.d.m.

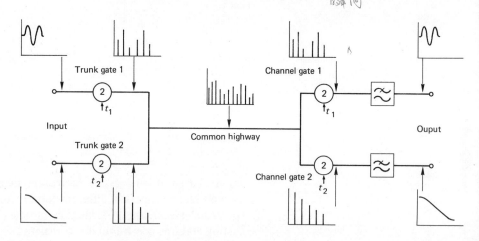

Fig. 2.2 A simple 2-channel t.d.m. system. t_1 = a series of pulses occurring at fixed intervals. t_2 = a series of pulses occurring at the same periodicity as t_1 but commencing later by an amount equal to half the time interval (From the *Post Office Electrical Engineers Journal*)

system is illustrated by Fig. 2.2 which shows the basic arrangement of a two-channel t.d.m. system.

The two channels which are to share the common circuit are each connected to it via a CHANNEL GATE. The channel gates are electronic switches which only permit the signal present on a channel to pass when opened by the application of a controlling pulse. Hence, if the controlling pulse is applied to gate 1 at time t_1 and not to gate 2, gate 1 will open for a time equal to the duration of the pulse but gate 2 will remain closed. During this time, therefore, a pulse or sample of the amplitude of the signal waveform on channel 1 will be transmitted to line. At the end of the pulse, both gates are closed and no signal is transmitted to line. If now the controlling pulse is applied to gate 2 at a later time t_2, gate 2 will open and a sample of the signal waveform on channel 2 will be transmitted. Thus if the pulses applied to control the opening and shutting of gates 1 and 2 are repeated at regular intervals, a series of samples of the signal waveforms existing on the two channels will be transmitted.

At the receiving end of the system gates 1 and 2 are opened, by the application of control pulses, at those instants when the incoming waveform samples appropriate to their channel are being received. This requirement demands accurate SYN-CHRONIZATION between the controlling pulses applied to the gates 1, and also between the controlling pulses applied to the gates 2. If the time taken for signals to travel over the common circuit was zero then the system would require controlling pulses as shown, but since, in practice, the transmission time is not zero, the controlling pulses applied at the receiving end of the system must occur slightly later than the corresponding controlling pulses at the sending end. If the pulse synchronization is correct the waveform samples are directed to the correct channels at the receiving end. The received samples must then be converted back to the original waveform, i.e. demodulated. Provided the sampling rate, i.e. the number of controlling pulses per second, is at least equal to twice the highest frequency contained in the original waveform, correct demodulation can be achieved by merely passing the samples through a filter network that passes freely all frequencies lower than the sampling frequency.

In the foregoing description of a simple t.d.m. system it has been assumed that the samples of the information signal are directly transmitted to line. Very often, however, it is desirable for the transmitted pulses to occupy an entirely different part of the frequency spectrum. For example with radio-frequency pulses the samples may be efficiently radiated by an aerial. When this is the case the equipment required at each end of a t.d.m. system is somewhat more complex than that shown in the diagram.

Types of Modulation

For a signal to be frequency translated to another part of the frequency spectrum it is necessary for the signal to vary one of the characteristics of a sinusoidal wave—usually known as the CARRIER WAVE—whose frequency occupies the required part of the spectrum. The process by which one of the characteristics of a carrier wave is modified in accordance with the characteristics of a modulating signal is known as modulation.

The general expression for a sinusoidal carrier wave is

$$v = V \sin(\omega t + \theta) \qquad (2.1)$$

where v = instantaneous voltage of the wave,
V = peak value or amplitude of the wave,
ω = angular velocity of the wave in radians/second; ω is related to the frequency of the wave by the expression $\omega = 2\pi f$ where f is the frequency in hertz,
θ = the phase of the wave at the instant when $t = 0$.

For modulation it is necessary to cause one of the characteristics of the wave to be varied in accordance with the waveform of the modulating signal. Three possibilities exist:

(a) the amplitude V of the wave may be varied to give amplitude modulation,
(b) the frequency $\omega/2\pi$ may be modified to give frequency modulation,
(c) the phase θ may be the variable, giving phase modulation.

To illustrate, graphically, the difference between these three types of modulation consider the case in which the modulating signal is a positive-going pulse as in Fig. 2.3a.

In the case of the *amplitude-modulated* wave, the frequency of the carrier wave remains constant but its amplitude is suddenly increased when the leading edge of the modulating pulse occurs and then suddenly reduced to its original value when the pulse ends (Fig. 2.3c). The *phase-modulated* carrier wave has a constant amplitude and frequency, but experiences an abrupt change of phase at the instants corresponding to the beginning and end of the modulating signal pulse (Fig. 2.3d). Finally, the *frequency-modulated* carrier wave also has a constant amplitude, but the frequency of the wave is abruptly increased at the beginning of the modulating pulse and maintained at the new frequency until the pulse ends, when the frequency is suddenly decreased to its original value (Fig. 2.3e).

Amplitude modulation is the most commonly used method in Great Britain. It is employed for f.d.m. telephony and telegraphy systems, radio broadcasting in the long, medium and short wavebands, the picture signals of both 405 line and

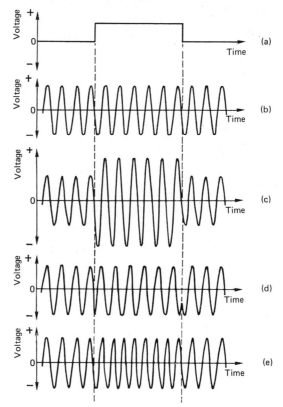

Fig. 2.3 Illustrating the difference between amplitude, frequency, and phase modulation
(a) Modulating signal
(b) Unmodulated carrier wave
(c) Amplitude-modulated wave
(d) Phase-modulated wave
(e) Frequency-modulated wave

625 line television broadcasting, the sound signals of 405 line television broadcasting and for various point-to-point and base-to-mobile radio-telephony systems.

Frequency modulation is used for v.h.f. radio broadcasting and for the sound signal of 625 line television broadcasting, as well as for some base-to-mobile radio-telephony systems. Phase modulation is employed in its own right in data transmission links but it is mainly used as a stage in the production of a frequency-modulated signal. Frequency modulation has the advantage over amplitude modulation in that its performance in the presence of interfering signals and noise can be much better. It suffers from the disadvantage of requiring a larger bandwidth. Because of the large frequency band required for a frequency-modulated system the use of frequency modulation for broadcast and telephony circuits is restricted to very high frequencies where the bandwidth required can be more readily provided.

Systems using the principle of time-division multiplex are generally based on the sampling of the amplitude of the information signal at regular intervals, and the subsequent

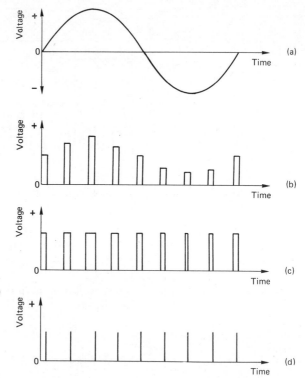

Fig. 2.4 Illustrating the difference between p.a.m., p.d.m. and p.p.m.
(a) Modulating signal
(b) Pulse-amplitude modulation
(c) Pulse-duration modulation
(d) Pulse-position modulation

transmission of a short pulse of radio-frequency carrier for each sample. For the intelligence contained in the information signal to be transmitted, the pulse characteristics must, in some way, be varied in accordance with the amplitude of the signal at the instant of sampling. Four main types of pulse modulation exist, namely pulse-amplitude modulation (p.a.m.), pulse-duration modulation (p.d.m.), pulse-position modulation (p.p.m.) and pulse-code modulation (p.c.m.). The principle of each of the first three types of pulse modulation is shown in Fig. 2.4. The fourth type will be discussed in Chapter 8.

With P.A.M., pulses of radio-frequency carrier of equal width and common spacing are employed and have their amplitudes varied in proportion to the instantaneous amplitude of the modulating signal. P.D.M. employs pulses of constant amplitude and common spacing between their leading edges, but whose width, or duration, is varied in accordance with the waveform of the modulating signal. Finally P.P.M. employs constant amplitude and width pulses whose position (in time) is dependent upon the instantaneous amplitude of the modulating signal.

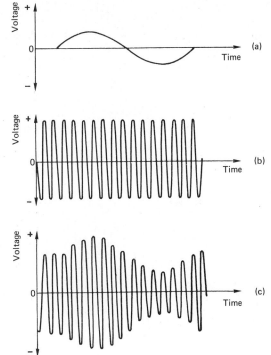

Fig. 2.5 A carrier wave modulated by a sinusoidal signal
(a) Modulating signal
(b) Unmodulated carrier wave
(c) Amplitude-modulated carrier wave

Each of these three methods of pulse modulation suffers from the disadvantage that distortion and noise is cumulative along the system—although the p.p.m. system is much less affected than the other two in this way. The fourth type of pulse modulation, i.e. pulse code modulation (p.c.m.), is a more sophisticated method which largely overcomes the effects of distortion and noise, but at the expense of added complexity and a wider bandwidth.

Amplitude Modulation

If a carrier wave is amplitude modulated, the amplitude of the carrier is caused to vary in accordance with the instantaneous value of the modulating signal. For example, consider the most simple case when the modulating signal is itself sinusoidal. The amplitude of the modulated wave must then vary in a sinusoidal manner as shown in Fig. 2.5. The difference in frequency between the carrier wave and the modulating signal will normally be much higher than is indicated by the diagram. In practice, the frequency difference would often be of the order of thousands of hertz.

26 MODULATION

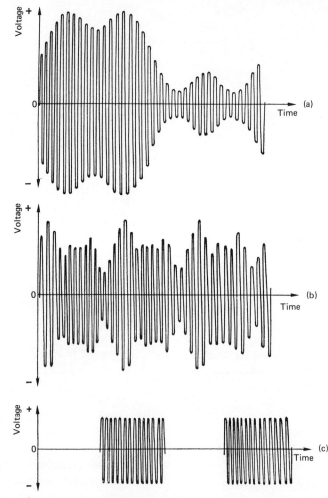

Fig. 2.6 Waveforms of a carrier wave modulated by
(a) a fundamental plus third harmonic
(b) a cello organ pipe playing C
(c) a telegraphy signal

The amplitude of the modulated carrier wave can be clearly seen to vary between a maximum value, which is greater than the amplitude of the unmodulated wave, and a minimum value, which is less than the amplitude of the unmodulated wave. The outline of the modulated carrier waveform is known as the modulation ENVELOPE.

To convey intelligence, a signal containing at least two components at different frequencies is required and hence, in practice, a sinusoidal modulating signal is rarely used. However, whatever the waveform of the modulating signal the same principle still holds good: the envelope of the modulated wave must be the same as the waveform of the modulating signal. For example, Fig. 2.6 shows the envelopes of a carrier wave modulated by (a) a signal consisting of components at a fundamental frequency and its third harmonic, both components being in phase at time $t = 0$, (b) note **C** played on a cello organ pipe and (c) a telegraphy signal produced by a teleprinter.

The Frequencies in an Amplitude-Modulated Wave

It is possible to show, with the aid of mathematics beyond the scope of this book, that when a sinusoidal carrier wave is amplitude modulated, each frequency component in the modulating signal gives rise to two frequencies in the modulated wave, one below the carrier frequency and one above. When, for example, the modulating signal is a sinusoidal wave of frequency f_m the modulated carrier wave contains three frequencies:

(a) the carrier frequency f_c.
(b) the "lower sidefrequency" $(f_c - f_m)$.
(c) the "upper sidefrequency" $(f_c + f_m)$.

The two new frequencies are the upper and lower SIDE-FREQUENCIES $(f_c + f_m)$ and $(f_c - f_m)$ respectively, and these are equally spaced either side of the carrier frequency by an amount equal to the modulating signal frequency f_m. The frequency f_m of the modulating signal itself is *not* present.

The bandwidth required to transmit an amplitude modulated carrier wave is equal to the difference between the highest frequency to be transmitted and the lowest. In the case of sinusoidal modulation and with $f_c > f_m$, the bandwidth required is given by

$$B = (f_c + f_m) - (f_c - f_m) = 2f_m$$

i.e. the required bandwidth is equal to twice the frequency of the modulating signal.

EXAMPLE 2.1

A 100 kHz carrier is amplitude modulated by a sinusoidal tone of frequency 3000 Hz. Determine the frequencies contained in the amplitude-modulated wave and the bandwidth required for its transmission.

Solution
The frequencies contained in the modulated wave are

(1) The carrier frequency $f_c = 100\,000$ Hz   (*Ans.*)
(2) The lower sidefrequency $(f_c - f_m) = 100\,000 - 3000$
$= 97\,000$ Hz   (*Ans.*)
(3) The upper sidefrequency $(f_c + f_m) = 100\,000 + 3000$
$= 103\,000$ Hz   (*Ans.*)
The bandwidth required $= 103\,000 - 97\,000 = 6000$ Hz   (*Ans.*)

If the modulating signal is non-sinusoidal it will contain components at a number of different frequencies; suppose that the highest frequency contained in the modulating signal is f_2 and the lowest frequency is f_1. Then the frequency f_2 will produce an upper sidefrequency component (f_c+f_2) and a lower sidefrequency component (f_c-f_2), while frequency f_1 will produce upper and lower sidefrequencies $(f_c \pm f_1)$. Thus the modulating signal will produce a number of lower sidefrequency components lying in the range (f_c-f_2) to (f_c-f_1) and a number of upper sidefrequency components in the range (f_c+f_1) to (f_c+f_2). The band of frequencies below the carrier frequency, i.e. (f_c-f_2) to (f_c-f_1), is known as the LOWER SIDEBAND, while the band of frequencies above the carrier frequency is known as the UPPER SIDEBAND. When the carrier frequency is higher than the modulating signal frequency the sidebands are symmetrically situated, with respect to frequency, on either side of the carrier frequency. The lower sideband is said to be *inverted* because the highest frequency in it (f_c-f_1) corresponds to the lowest frequency f_1 in the modulating signal and vice versa. Similarly, the upper sideband is said to be *erect* because the lowest frequency in it (f_c+f_1) corresponds to the lowest frequency f_1 in the modulating signal.

EXAMPLE 2.2

A 108 kHz carrier wave is amplitude modulated by a band of frequencies, 300–3400 Hz. What frequencies are contained in the upper and lower sidebands of the a.m. wave and what is the bandwidth required to transmit the wave?

Solution

The lower sideband will contain frequencies in the band

108 000 − 3400 Hz to 108 000 − 300 Hz,
or 104 600 to 107 700 Hz   (*Ans.*)

The upper sideband will contain frequencies between

108 000 + 300 Hz to 108 000 + 3400 Hz
or 108 300 to 111 400 Hz   (*Ans.*)

The required bandwidth B is equal to the maximum frequency contained in the modulated wave minus the minimum frequency:

$B = 111\,400 - 104\,600 = 6800$ Hz   (*Ans.*)

Note that the bandwidth is equal to twice the highest frequency contained in the modulating signal.

It is possible to confirm graphically the presence of a number of frequencies in an amplitude-modulated wave. Consider, for example, the case of a 10 000 Hz carrier wave and a 2000 Hz modulating signal. The upper sidefrequency will be 12 000 Hz and the lower sidefrequency will be 8000 Hz. The

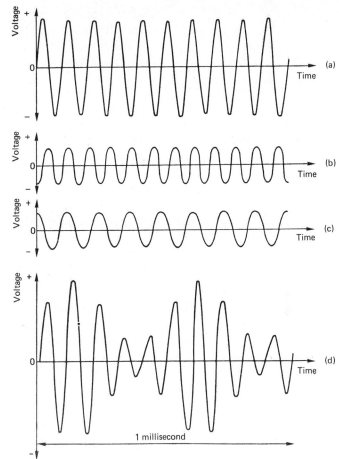

Fig. 2.7 Showing the formation of an amplitude-modulated wave by adding the components at the carrier and upper and lower sidefrequencies.
(a) Carrier wave
(b) Upper sidefrequency
(c) Lower sidefrequency
(d) Amplitude-modulated carrier wave

waveforms of the carrier, the upper sidefrequency and the lower sidefrequency components are shown in Figs. 2.7a, b and c respectively. The carrier component can be seen to complete 10 cycles, the upper sidefrequency component 12 cycles and the lower sidefrequency component 8 cycles in the time of one millisecond. To obtain the waveform of the amplitude-modulated wave, which is composed of these three components, it is necessary to sum the instantaneous values of the three waves. It can be seen that the envelope of the modulated wave shows two complete variations in the time of one millisecond, i.e. it varies at the modulating frequency of 2000 Hz.

There are two ways in which the frequency spectrum occupied by a modulated carrier wave may be shown. Each component may be represented by an arrow drawn perpendicular to the frequency axis as shown in Fig. 2.8, where f_c, f_1 and f_2 have the same meanings as before. The lengths of the

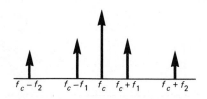

Fig. 2.8 Frequency spectrum of an amplitude-modulated wave

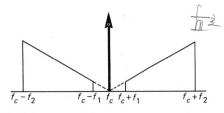

Fig. 2.9 A method of representing the sidebands of amplitude modulation

arrows are made proportional to the amplitude of the component they each represent. If many of the frequencies between f_1 and f_2 were present in the modulating signal, a large number of arrows would be required and the diagram would become impracticable. It is usual, particularly in connection with multi-channel carrier telephony systems, to represent the sidebands produced by a complex modulating signal by truncated triangles, in which the vertical ordinates are made proportional to the modulating frequency and no account is taken of amplitude—see Fig. 2.9. This method of representing sidebands gives an immediate indication of which sideband is erect and which is inverted. This is useful when considering systems employing more than one stage of modulation when the inverted sideband is not necessarily the lower sideband.

Modulation Depth

The envelope of an amplitude-modulated carrier wave varies in accordance with the waveform of the modulating signal and hence there must be a relationship between the maximum and minimum values of the modulated wave and the amplitude of the modulating signal. This relationship is expressed in terms of the modulation factor of the modulated wave.

The MODULATION FACTOR m of an amplitude-modulated wave is defined by the expression

$$m = \frac{\text{maximum amplitude} - \text{minimum amplitude}}{\text{maximum amplitude} + \text{minimum amplitude}} \quad (2.2)$$

When expressed as a percentage m is known as the modulation depth, or the depth of modulation or the percentage modulation.

Consider a sinusoidally modulated wave such as the wave shown in Fig. 2.10. Since the envelope of the modulated carrier wave must vary in accordance with the modulating signal, its maximum amplitude must be equal to the amplitude of the carrier wave plus the amplitude of the modulating signal, i.e. $(V_c + V_m)$. Similarly, the minimum amplitude of the modulated wave must be equal to $(V_c - V_m)$, where V_c is the amplitude of the carrier wave and V_m is the amplitude of the modulating signal.

For *sinusoidal modulation*, therefore, the modulation factor m becomes

$$m = \frac{(V_c + V_m) - (V_c - V_m)}{(V_c + V_m) + (V_c - V_m)} = \frac{V_m}{V_c} \quad (2.3)$$

i.e. the modulation factor is equal to the ratio of the amplitude of the modulating signal to the amplitude of the carrier wave.

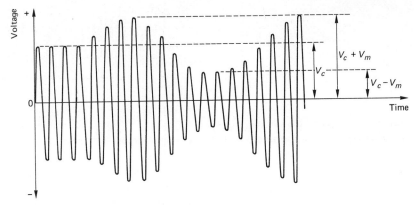

Fig. 2.10 Characteristics of an a.m. wave required to calculate modulation depth

It can be shown, mathematically, that for a sinusoidally modulated carrier wave, the amplitudes of the two sidefrequencies are the same and equal to $\tfrac{1}{2}m$ times the amplitude of the carrier wave.

EXAMPLE 2.3

Draw the waveform of an amplitude-modulated carrier wave that is sinusoidally modulated to a depth of 25%.

Solution
For a modulation depth of 25%, $m = 0.25$, and since the modulating signal is sinusoidal, $m = V_m/V_c$ or $V_m = 0.25 V_c$. Thus

the maximum amplitude of the modulated wave is $V_c + 0.25 V_c = 1.25 V_c$ and
the minimum amplitude is $V_c - 0.25 V_c = 0.75 V_c$

The required waveform is shown in Fig. 2.11.

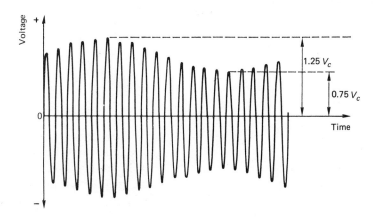

Fig. 2.11 Amplitude-modulated wave of modulation depth 25%

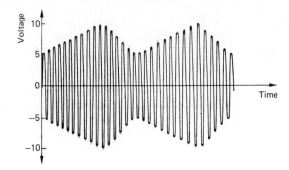

Fig. 2.12 Carrier wave amplitude-modulated by a non-sinusoidal waveform

EXAMPLE 2.4

Determine the depth of modulation of the amplitude-modulated wave shown in Fig. 2.12.

Solution
The maximum voltage of the wave is 10 V and the minimum voltage is 5 V; hence the depth of modulation is

$$\frac{10-5}{10+5} \times 100\% = 33.3\% \quad (Ans.)$$

EXAMPLE 2.5

The envelope of a sinusoidally modulated carrier wave varies between a maximum value of 8 V and a minimum value of 2 V. Find (*a*) the amplitude of the carrier frequency component, (*b*) the amplitude of the modulating signal and (*c*) the amplitude of the two sidefrequencies.

Solution

(*a*) Maximum value $V_c + V_m = 8$ (2.4)
Minimum value $V_c - V_m = 2$ (2.5)

Adding equations (2.4) and (2.5)

$2V_c = 10$ or $V_c = 5$ V (*Ans.*)

(*b*) Subtracting equation (2.5) from equation (2.4)

$2V_m = 6$ or $V_m = 3$ V (*Ans.*)

(*c*) The amplitude of each of the two sidefrequencies is given by

$$\tfrac{1}{2}mV_c \quad \text{i.e.} \quad \frac{V_m}{V_c} \times \frac{V_c}{2} \quad \text{or} \quad \tfrac{1}{2}V_m$$

Amplitude of sidefrequencies = 3/2 or 1.5 V (*Ans.*)

When the depth of modulation is 100% the modulation envelope varies between a maximum of $2V_c$ and a minimum of 0. If the depth of modulation is increased beyond this value the modulation envelope becomes distorted as shown in Fig. 2.13.

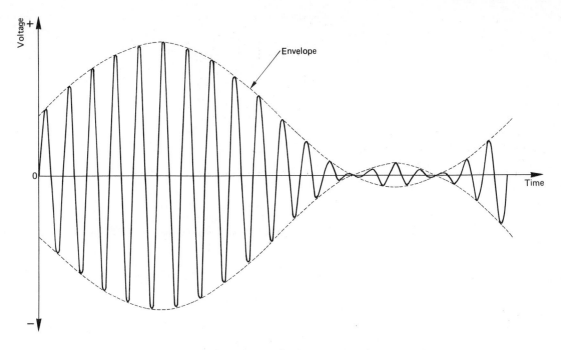

Fig. 2.13 Amplitude-modulated wave with a modulation depth in excess of 100%

Since the envelope of the modulated wave is no longer a replica of the modulating signal waveform, considerable distortion has taken place. This, of course, means that modulation depths greater than 100% are never employed in practice.

Power of an Amplitude-modulated Wave

When an amplitude-modulated voltage wave is applied across a resistance R, each component frequency of the wave will dissipate power. The total power dissipated by the wave is the sum of the powers dissipated by the individual components. In the case of a sinusoidally modulated carrier wave, three components exist; namely, the carrier frequency, the upper sidefrequency, and lower sidefrequency.

The power P_c developed by each of the sidefrequency components is

$$P_c = \left(\frac{V_c}{\sqrt{2}}\right)^2 \cdot \frac{1}{R} = \frac{V_c^2}{2R} \text{ watts}$$

and the power developed by each of the sidefrequency components is

$$P_{LSF} = P_{USF} = \left(\frac{V_m}{2\sqrt{2}}\right)^2 \cdot \frac{1}{R} = \left(\frac{mV_c}{2\sqrt{2}}\right)^2 \cdot \frac{1}{R} = \frac{m^2 V_c^2}{8R} \text{ watts}$$

so that the total power P_t is

$$P_t = \frac{V_c^2}{2R} + \frac{m^2 V_c^2}{8R} + \frac{m^2 V_c^2}{8R}$$

or

$$P_t = \frac{V_c^2}{2R}\left(1 + \frac{m^2}{2}\right) \text{ watts} \tag{2.6}$$

EXAMPLE 2.6

The total power dissipated by an amplitude-modulated wave is 1575 W. Calculate the power in the sidefrequencies if the modulation depth is (i) 50% and (ii) 100%.

Solution
From equation (2.6), where P_c is the carrier power,

$1575 = P_c(1 + \tfrac{1}{2}m^2)$

(i) When $m = 0.5$

$1575 = P_c(1 + \tfrac{1}{2}0.25)$

$P_c = 1575/1.125 = 1400$ W

Therefore $P_{SF} = 1400 \times 0.125 = 175$ W   (*Ans.*)

(ii) When $m = 1$

$1575 = P_c(1 + \tfrac{1}{2})$

$P_c = 1575/1.5 = 1050$ W

Therefore $P_{SF} = 1050 \times 0.5 = 525$ W   (*Ans.*)

It is clear from this example that the power contained in the two sidefrequencies is only a small fraction of the total power, rising to a maximum when $m = 1$, that is only one-third of the total power. Since it is only the sidefrequencies that carry information, amplitude modulation is not a very efficient system when considered on a power basis.

Signal-to-Noise Ratio

The output of a communication system, whether line or radio, will always contain some unwanted voltages or currents in addition to the desired signal. The unwanted output signal is known as *noise* and may have one or more of a number of different causes. For the signal received at the end of a system to be of use, the signal power must be greater than the noise power by an amount depending upon the nature of the signal.

The ratio of the wanted signal power to the unwanted noise power is known as the *signal-to-noise ratio*. The signal level must never be allowed to fall below the value that gives the required minimum signal-to-noise ratio because any gain introduced thereafter will increase the level of both noise and signal by the same amount and will not improve the signal-to-noise ratio.

EXAMPLE 2.7

The signal voltage at the output of an amplifier is 0.866 V and the noise voltage that is unavoidably also present is 10 mV. Calculate the signal-to-noise ratio at the output of the amplifier.

Solution
Signal-to-noise ratio is the ratio of the wanted signal *power* to the unwanted noise *power*. Therefore

$$\text{Signal-to-noise ratio} = \frac{(0.866)^2}{(10 \times 10^{-3})^2} = 7500 \quad (Ans.)$$

Noise may arise from a number of different sources, some natural and some man-made (motor car ignition systems and electric motors for example). Most naturally occurring noise sources produce a noise power which is directly proportional to bandwidth. This means that the signal-to-noise ratio at the output of an amplifier, a radio receiver, or some other kind of electronic, line or radio equipment is inversely proportional to the bandwidth of that equipment.

EXAMPLE 2.8

The signal-to-noise ratio at the output of a radio-frequency amplifier is 1000. What would be the signal-to-noise ratio if the bandwidth of the amplifier were doubled?

Solution
If the bandwidth of the amplifier is doubled the noise power at the output terminals will also be doubled. The signal output power is unchanged and so the signal-to-noise ratio will be reduced by half to 500.

Single-sideband Operation

It is clear that an amplitude-modulated wave contains the intelligence represented by the modulating signal in both the upper sideband and the lower sideband. It is therefore unnecessary to transmit both sidebands. Furthermore, the carrier component is of constant amplitude and frequency and does not carry any information. It is possible (using a balanced modulator and a filter) to suppress both the carrier and one sideband in the transmitting equipment, and to transmit just

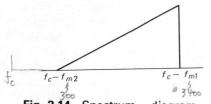

Fig. 2.14 Spectrum diagram of s.s.b. signal

the other sideband without any loss of information. This method of operation is known as *single-sideband suppressed-carrier* (s.s.b.) working. The frequency spectrum graph of an s.s.b. signal is shown in Fig. 2.14. S.S.B. operation of a communication system offers the following advantages over double-sideband (d.s.b.) working.

(a) The bandwidth required for s.s.b. transmission is only half as great as that required for d.s.b. transmission carrying the same information. This allows more channels to be operated within the frequency spectrum provided by the transmission medium.

(b) The signal-to-noise ratio at the receiving end of an s.s.b. system is greater than that of a d.s.b. system. The improvement is 9dB for a depth of modulation of 100% and even more for modulation depths of less than 100%; some of this improvement is the result of an increase in the ratio sideband-power/total-power of the transmitted output and the rest is because the necessary bandwidth is reduced by half (noise power is proportional to bandwidth).

(c) A d.s.b. transmitter produces a power output (due to the transmitted carrier) at all times, whereas an s.s.b. transmitter does not. A saving in the consumption of the d.c. power taken from the power supply is thus obtained, with an overall increase in transmitter efficiency.

(d) Radio waves are subject to a form of interference known as *selective fading*. When this is prevalent, considerable distortion of a d.s.b. signal may occur, because the carrier component may fade below the level of the sidebands, so that the two sidebands beat together to produce a large number of unwanted frequencies. This cannot occur with an s.s.b. system because the signal is demodulated against a locally generated carrier of constant amplitude.

(e) In multi-channel telephony line systems any non-linearity produces intermodulation products, many of which would lead to inter-channel crosstalk. Most non-linearity arises in the output stages of the line amplifiers since it is these that handle the signals of the greatest amplitude. Suppression of the carrier component reduces the signal levels to be handled by the amplifiers and this limits the effect of any non-linearity and hence also reduces crosstalk.

The disadvantage of s.s.b. working is the need for more complex, and hence more costly, receiving equipment. The increased complexity is due to the need to reintroduce a carrier at the same frequency as the carrier originally suppressed at the transmitter. Any lack of synchronization between the suppressed and reintroduced carrier frequencies produces a shift in each component frequency of the demodulated signal. To preserve intelligibility the reinserted carrier must be within a few cycles of the original carrier frequency. The extra cost and complexity of operation are the reasons why s.s.b. working is restricted to long-distance radio and line telephony systems and is not employed for domestic radio broadcasting.

Frequency Modulation

If a carrier wave is frequency-modulated, its frequency is made to vary in accordance with the instantaneous value of the modulating signal. The amount by which the carrier frequency deviates from its nominal value is proportional to the amplitude of the modulating signal, and the number of times per second the carrier deviates is equal to the modulating frequency. Fig. 2.15b shows a frequency-modulated wave and Fig. 2.15a the corresponding modulating signal. Over the time interval 0 to t_1 the modulating signal voltage is zero and so the carrier is unmodulated. From t_1 to t_2 the modulating signal

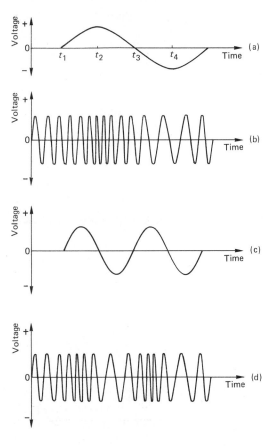

Fig. 2.15 Frequency-modulated waves

voltage is increasing in the positive direction and the carrier frequency increases to a maximum value at t_2. After time t_2 the modulating voltage falls towards zero, and the carrier frequency reduces in value until at time t_3 it has reached its unmodulated value. During the following negative half cycle of the modulating signal voltage, the carrier frequency is reduced below its unmodulated value. The carrier frequency has its minimum value when the modulating signal voltage reaches its peak negative value, i.e. at time t_4.

If the frequency of the modulating signal is increased (Fig. 2.15c), the number of times per second that the carrier frequency is varied about its mean, unmodulated value is increased in proportion. This is shown by Fig. 2.15d. It should be noticed that the minimum and maximum values of the modulated carrier frequency are the same in both Figs. 2.15b and d; this is because the respective modulating signals have equal amplitudes.

Since the amplitude of a carrier wave is not changed when it is frequency modulated, there is no change in the power carried by the wave.

Frequency Deviation

The frequency deviation of a frequency-modulated wave is the amount by which the carrier frequency has been changed from its unmodulated value. Frequency deviation is proportional to the amplitude of the modulating signal voltage. There is no inherent maximum to the frequency deviation that can be achieved (unlike amplitude modulation where 100% modulation is the maximum amplitude deviation possible). Since the bandwidth occupied by a frequency-modulated wave is proportional to frequency deviation, a maximum value to the deviation permitted in a particular system must be arbitrarily chosen. The maximum value of frequency deviation that is allowed to occur in a particular system is known as the *rated system deviation*. Since frequency deviation is proportional to modulating signal amplitude, it follows that a maximum value, which depends on the sensitivity of the modulator, is also determined for the modulating voltage.

EXAMPLE 2.9

The rated system deviation in the BBC v.h.f. sound broadcast system is 75 kHz. What will be the frequency deviation of the carrier if the modulating signal voltage is (i) 50% and (ii) 20% of the maximum value permitted?

Solution
(i) Frequency deviation = 75 kHz × 0.5 = 37.5 kHz   (*Ans.*)
(ii) Frequency deviation = 75 kHz × 0.2 = 15 kHz   (*Ans.*)

The term *frequency swing* is sometimes applied to a frequency-modulated wave. It refers to the maximum carrier frequency minus the minimum carrier frequency, i.e. the frequency swing is equal to twice the frequency deviation.

Modulation Index

When a carrier voltage is frequency modulated, its phase is also caused to vary. The peak phase deviation produced is equal to the ratio of the frequency deviation to the modulating frequency and is known as the modulation index of the modulated wave.

EXAMPLE 2.10

A carrier is frequency modulated by a 10 kHz sinusoidal wave whose amplitude is one-half the maximum permitted value. If the rated system deviation is 50 kHz determine (i) the frequency deviation and (ii) the phase deviation, of the carrier produced.

Solution
(i) Frequency deviation $= 50 \text{ kHz} \times 0.5 = 25 \text{ kHz}$ (*Ans.*)

(ii) Phase deviation = Modulation index $= \dfrac{\text{Frequency deviation}}{\text{Modulating frequency}}$ (2.7)

$= \dfrac{25 \text{ kHz}}{10 \text{ kHz}} = 2.5 \text{ rad}$ (*Ans.*)

When both the frequency deviation and the modulating frequency are at their maximum permitted values, the modulation index is then known as the *deviation ratio*.

The Frequencies in a Frequency-modulated Wave

When a carrier of frequency f_c is frequency modulated by a sinusoidal wave of frequency f_m, the resultant waveform contains components at a number of different frequencies. The modulated wave contains the following frequency components:

(*a*) the carrier frequency f_c
(*b*) first-order sidefrequencies $f_c \pm f_m$
(*c*) second-order sidefrequencies $f_c \pm 2f_m$
(*d*) third-order sidefrequencies $f_c + 3f_m$

and so on. The number of sidefrequencies present in a particular wave depends upon its modulation index, the larger the value of the modulation index the greater the number of sidefrequencies generated. The amplitudes of the various components, including the carrier itself, vary in a complicated manner as the modulation index is increased. Any component, again including the carrier, may have zero amplitude at a particular value of modulation index.

The bandwidth required to transmit a frequency-modulated wave is not, as might be expected, twice the frequency deviation of the carrier but, instead, is greater. The expression generally used to determine the bandwidth occupied by a frequency-modulated wave is given by equation (2.8), i.e.

$$\text{Bandwidth} = 2(f_{dev} + f_m) \tag{2.8}$$

where f_{dev} = frequency deviation of carrier, f_m = modulating frequency.

EXAMPLE 2.11

The BBC v.h.f. frequency-modulated sound transmissions are operated with a rated system deviation of 75 kHz and a maximum modulating frequency of 15 kHz. Determine the required bandwidth. Compare this with the bandwidth required by an amplitude-modulation system that provides the same audio bandwidth.

Solution
From equation (2.8),

$$\text{Bandwidth} = 2(75 \times 10^3 + 15 \times 10^3) = 180 \text{ kHz} \quad (Ans.)$$

If the same audio bandwidth were to be provided by an amplitude-modulated system, the necessary r.f. bandwidth would be $f_c \pm 15$ kHz or 30 kHz. Clearly, in this case, frequency modulation is much more expensive in its use of the available frequency spectrum than is amplitude modulation. The signal-to-noise ratio at the output of a frequency-modulated system is proportional to the deviation ratio and, since D = maximum-frequency-deviation/maximum-modulating-frequency, also to the bandwidth occupied. This means that the signal-to-noise ratio can always be improved at the cost of an increased bandwidth; this is the reason why the BBC frequency-modulated sound broadcast stations are operated in the v.h.f. band, since here the necessary wide bandwidth is available. With narrow-band frequency modulation, the deviation ratio is small and the system offers little, if any, improvement in signal-to-noise ratio over the use of amplitude modulation.

The Relative Merits of Frequency Modulation and D.S.B. Amplitude Modulation

Frequency modulation offers the following advantages over the use of d.s.b. amplitude modulation:

(*a*) The signal-to-noise ratio at the output of the f.m. receiver can be greater than that of a d.s.b. amplitude-modulation receiver.

MODULATION

(b) The amplitude of a frequency-modulated wave is constant, allowing more efficient transmitters to be built.

(c) An f.m. receiver possesses the ability to suppress the weaker of two signals simultaneously received at or near the same frequency. This effect is known as the *capture effect*.

(d) The dynamic range, i.e. the range of modulating signal amplitudes that can be transmitted, is much larger.

The disadvantage of frequency modulation is the (generally) wider bandwidth required.

Amplitude modulation is used for long, medium, and short waveband broadcast transmissions, for the picture signal in television broadcasting, for long-distance radio-telephony, for v.h.f./u.h.f. base-mobile systems, and for various ship and aeroplane radio services. Frequency modulation is used for v.h.f. sound broadcasting, for the sound signal of u.h.f. television broadcasts, for some base-mobile systems, for some ship/aero services, and for wideband radio-telephony systems.

Data Transmission

Nowadays digital computers are widely used by many organizations for scientific and engineering computations, and for commercial data processing applications, such as the calculation and addressing of bills and the compilation of company statistics and records. Much of the data to be processed and perhaps stored by a computer is originated at, and the results required by, offices and laboratories that are not located at the same geographical point as the computer installation. This means that there is an ever-increasing demand for data links that will enable branch offices to communicate with a central computer. The basic arrangement of a data system is shown in Fig. 2.16. Communication with the computer is carried out via the keyboard of a teleprinter or via a card- or tape-reader since a computer needs its input data in binary form. More

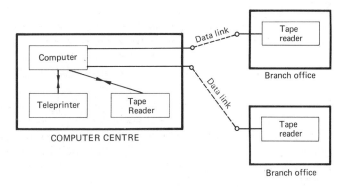

Fig. 2.16 Communication with a computer

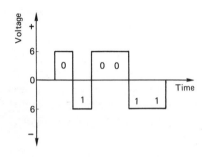

Fig. 2.17 Data waveform

characters are required than for a telegraphy system and further error-checking *parity bits* are added. The codes used for data transmission therefore use more than the $7\frac{1}{2}$ units that have long been standardized for telegraphy.

Access to the computer will be possible from several different points within the computer centre and also from a number of branch offices linked by telephone line to the centre. Sometimes more than one computer may be integrated within the data network of a particular organization and in such cases there will also be a need for two computers to be able to communicate directly with one another.

The data fed into a computer is in binary form with the binary number 0 being represented by a positive 6 V voltage and binary 1 by −6 V (see Fig. 2.17). Each binary digit, 0 or 1, is known as a *bit*. The data signalling rate is the rate at which information can be transmitted and for the binary code this is equal to the reciprocal of the time duration of a bit. The data waveform of Fig. 2.17 consists of a d.c. component (equal to the average value of the waveform), a fundamental frequency, and a number of harmonics of the fundamental frequency. Such a waveform cannot be transmitted over a data link which has been set up over a connection in the switched telephone network. There are three reasons for this: (i) transmission bridges in the telephone exchanges will not allow the d.c. component to pass; (ii) many switched connections will be routed over one or more multi-channel telephony systems and these do not provide a d.c. path; (iii) a switched connection is likely to include one or more amplified audio-frequency junctions and/or trunks. The line amplifiers are unable to transmit the d.c. component since they incorporate capacitors and/or inductors as coupling components.

When there is sufficient traffic to economically justify it, many organizations rent private circuits from the Post Office to permanently connect some of their branch offices to their computer centre. All but the shortest of these will also be routed over amplified and/or multi-channel circuits and so will also be unable to transmit the d.c. component.

The transmission of data over any but the shortest of links therefore necessitates the use of modulation to shift the data waveform to a more convenient part of the frequency spectrum. Short links are operated over normal telegraph-type (very restricted bandwidth) circuits using ±80 V teleprinter signals at, usually, 110 bits/sec.

Not all of the frequency spectrum of a commercial quality speech circuit can be made available for data transmission because it is essential to avoid the frequencies at which various signalling and supervisory tones are provided. The available frequency spectrum is therefore 900 to 2100 Hz. The modulation methods used for data systems are: (1) amplitude modulation, (2) frequency-shift (a form of frequency modulation) and (3) phase modulation.

(1) Amplitude Modulation

The data waveform can be used to switch a carrier wave of appropriate frequency on and off to produce the waveform shown in Fig. 2.18a. Binary digit 1 is represented by the transmission of the carrier wave and binary digit 0 by the

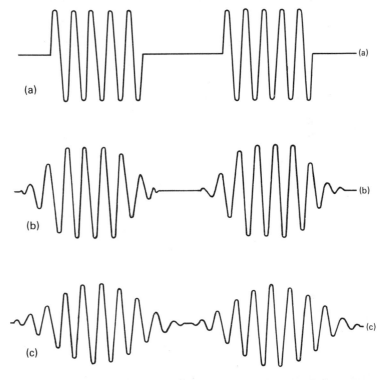

Fig. 2.18 Amplitude-modulated data waveforms

carrier being suppressed. Switching the carrier on and off in this way results in the generation of a large number of sidefrequencies and, to restrict the bandwidth occupied, the signal must be passed through a filter. Limiting the number of sidefrequencies transmitted increases the time taken for the envelope of the signal to reach its maximum value. The faster the data signalling rate the higher the fundamental frequency of the data waveform and the more widely spaced are the harmonics contained within that waveform. This means that the higher the signalling rate the more rounded the signal envelope will be; this is shown by Figs. 2.18b and c, in which waveform c corresponds to a higher signalling rate than waveform b. To reduce the occupied bandwidth still further, vestigial sideband operation is possible. Disadvantages of the method are that the accuracy of the received signal depends very much upon the signal-to-noise ratio of the link, and that any momentary breaks in the transmission path will be interpreted as binary 0 and errors will appear in the received data.

In practice, only single-sideband systems are employed and their use is restricted to high-speed computer-to-computer links operating at 48 kilobits/sec over wideband circuits in the frequency band 60–104 kHz.

(2) Frequency Shift

Frequency shift modulation of a carrier is a version of frequency modulation in which the carrier frequency can have one of two different values. One of these frequencies is used to represent the binary digit 0 while the other frequency stands for binary 1. The faster the data transmission rate required, the farther apart must be the two frequencies.

Data links connecting computer bureaux with their customers operate teleprinters at 110 bits/sec, or at 300 bits/sec, and use 1300 Hz to indicate binary digit 1 and 1700 Hz for binary 0. Data links operated as part of an organization's network use card and/or tape readers operating at either 600 bits/sec or 1200 bits/sec. The 600 bits/sec links use 1300 Hz and 1700 Hz to represent the binary digits 1 and 0, while the 1200 bits/sec links use 1300 Hz and 2100 Hz.

(3) Phase Modulation

It is possible to represent the two binary digits 0 and 1 by changing the phase of a constant-frequency carrier wave between two specified values. This method of modulation requires that the receiving end of the system has a reference wave of exactly the same frequency and of constant phase available, but this requirement involves considerable practical difficulties and is therefore rarely used. Most phase-modulated data systems employ differential phase modulation in which changes in phase, rather than the phase angle itself, are used to indicate the binary digits. In the present UK system the binary digits are grouped in pairs 00, 01, 10 and 11, and are indicated, respectively, by changes in phase of 0°, 90°, 180° and 270°.

Suppose for example that the binary data train 0100110010 is to be transmitted. Grouping the binary data train into pairs starting with the least significant digit gives 01, 00, 11, 00 and 10. The modulated carrier wave should therefore experience phase changes of, in turn, 90°, 0°, 270°, 0° and 180°. Differential phase modulation is used by higher speed data systems operating at 2400 bits/sec. The CCITT standard system, which is to be adopted in the UK uses a 45° phase shift to indicate the pair of digits 00; 135° phase shift to represent the digits 01; and phase shifts of 225° and 315° to signify 10 and 11 respectively.

Exercises

2.1. With reference to an amplitude-modulated wave what is meant by the terms sidefrequency, sideband and depth of modulation?

Sketch the waveform of an r.f. wave, amplitude-modulated by a sine-wave tone to a depth of 75 per cent.

If a radio-frequency carrier wave is amplitude-modulated by a band of frequencies, 300–3400 Hz, what will be the bandwidth of the transmission and what frequencies will be present in the transmitted wave if the carrier frequency is 104 kHz?
(C&G)

2.2. With reference to an amplitude-modulated wave, what is meant by the term sidebands?

Why do medium-wave broadcast receivers, used for the reception of amplitude-modulated signals, require a total bandwidth of about 9 kHz?

Briefly describe the effects on the reception of a medium-wave broadcast signal if the total bandwidth of the receiver were made (a) much less than, (b) much greater than, 9 kHz.
(C&G)

2.3. Sketch the waveforms of a radio-frequency carrier wave, amplitude-modulated by a sine-wave tone, when the depth of modulation is (a) 100 per cent, and (b) 25 per cent. If a radio-frequency carrier wave is amplitude-modulated by a band of speech frequencies, 50 Hz–4500 Hz, what will be the bandwidth of the transmission and what frequencies will be present in the transmitted wave, if the carrier frequency is 506 kHz?
(C&G)

2.4. What is an amplitude-modulated wave?

Why do broadcast receivers used for the reception of amplitude-modulated signals have a total bandwidth of about 9 kHz?
(C&G)

2.5. Sketch the waveform of a radio-frequency carrier wave amplitude-modulated by a sinusoidal tone when the depth of modulation is (a) 50% and (b) 100%. Label the axes clearly to show the quantities concerned.

If the amplitude of the unmodulated carrier is 1 volt (r.m.s.) and its frequency is 1000 kHz, what are the amplitudes and frequencies of the sidefrequency components in cases (a) and (b) when the modulation frequency is 1000 Hz? (C&G)

2.6. With reference to amplitude modulation explain the terms "modulation envelope" and "depth of modulation" and distinguish between "sidefrequencies" and "sidebands".

The amplitude of a 310 kHz wave is modulated sinusoidally at a frequency of 5 kHz between 0.9 V and 1.5 V. Determine the amplitude of the unmodulated carrier, the depth of modulation and the frequency components present in the modulated wave.
(C&G)

2.7. Draw, on graph paper, the envelope of a carrier wave amplitude-modulated to a depth of (i) 30%, (ii) 100% by a 2 kHz sinusoidal wave. The unmodulated carrier voltage is 10 V. Use scales of 1 V = 1 in. and 1 ms = 4 in. (or equivalent metric). Explain why a modulation depth greater than 100% would not be used in practice.

2.8. (a) A 100 kHz carrier is amplitude modulated, using a balanced modulator, by a 3 kHz signal and the lower sidefrequency is selected. This sidefrequency is then used to amplitude-modulate a 120 kHz carrier and again the lower sidefrequency is selected. Draw the frequency spectrum diagrams corresponding to the output of each modulator. Hence explain what is meant by the terms erect and inverted when applied to sidebands.

(b) List the advantages of s.s.b. amplitude modulation over d.s.b. amplitude modulation.

2.9. Explain, with the aid of suitable diagrams, what is meant by (i) frequency-division multiplex and (ii) time-division multiplex. Explain why such systems are used.

2.10. (a) A carrier frequency of 2 MHz is amplitude-modulated by a band of audio frequencies containing the harmonics of 50 Hz in the range up to 3 kHz. Sketch a frequency spectrum diagram to show the frequencies produced. (b) With reference to your diagram explain the difference between sidefrequencies and sidebands. (c) Determine the limits of the transmission in (a) above. (d) If the unmodulated carrier frequency of 2 MHz is propagated through a cable with a velocity of 2×10^8 m/s, determine its wavelength. (C&G)

2.11. (a) Sketch the waveform of an amplitude-modulated carrier signal sinusoidally modulated to a depth of 30%. Label the axes. (b) Use your diagram to explain what is meant by modulation envelope and depth of modulation. (c) The envelope of an amplitude-modulated waveform varies sinusoidally between maximum values of ±6 V and minimum values of ±2 V. Determine (i) the amplitude of the unmodulated carrier, (ii) the amplitude of the modulating signal, and (iii) the depth of modulation expressed as a percentage. (C&G)

2.12. Draw sketches to show the appearance of amplitude-, frequency- and phase-modulated waves when the modulating signal is of sawtooth waveform.

2.13. When a 1000 W carrier wave is amplitude-modulated by a sinusoidal wave the total power rises to 1200 W. Calculate (i) the power in the lower sidefrequency, and (ii) the depth of modulation. What would be the total power if the carrier were frequency modulated instead with a modulation index of 5?

2.14. Explain the meanings of the following terms used in conjunction with frequency modulation: (i) frequency deviation, (ii) rated system deviation, (iii) modulation index and (iv) deviation ratio.

The rated system deviation of the sound signals of u.h.f. television broadcasts is 50 kHz. What will be the frequency deviation if the modulating signal voltage is reduced to (i) 50% and (ii) 10% of its maximum permitted value?

2.15. Draw a frequency-modulated waveform. List the advantages of frequency modulation over amplitude modulation. Why is frequency modulation not used for medium-wave sound broadcasting? What is the necessary bandwidth for a frequency-modulated wave if the modulating frequency is 5 kHz and the frequency deviation is 20 kHz?

2.16. (a) Outline a method by which data may be transmitted over an audio circuit having a bandwidth of 300–3000 Hz. The required rate of transmission is 600 bits/sec.
(b) How could a simple method for detecting errors be incorporated in the system described in (a)? (C&G)

2.17. (a) One method of storing and transmitting data is by coding the information into binary form. (i) Discuss the advantages of coding in this form. (ii) How many different pieces of information may be coded using an eight-bit binary code?
(b) Outline a system of data transmission by which binary data may be transmitted over a distance of 10 km. State the maximum transmission rate of the suggested system. (C&G)

Short Exercises

2.18. What is meant by frequency modulation of a carrier wave? Why is it employed? Draw a frequency-modulated wave, making clear the type of modulating signal you have assumed.

2.19. What is the function of the carrier wave in a modulated system? Explain, with the aid of a frequency spectrum diagram, the production of a double-sideband amplitude-modulated signal.

2.20. Explain, with the aid of a frequency spectrum diagram, the production of a single sideband by removing the carrier and the other sideband.

2.21. Compare the relative merits of double- and single-sideband amplitude modulation, and frequency modulation.

2.22. Draw the waveform of a carrier wave which has been amplitude-modulated to a depth of about 120% and hence explain why such a modulation depth would not be employed in practice.

2.23. The frequency deviation of a frequency-modulated wave is 50 kHz. What does this statement mean? What is the value of the frequency swing? What is the modulation index when the modulating frequency is (i) 5 kHz and (ii) 10 kHz?

2.24. When an amplitude-modulated wave is applied across a 10 Ω resistor, the powers developed by the carrier and by each of the sidefrequency components are 100 W and 10 W respectively. Calculate the total power dissipated.

2.25. The signal voltage at the 75 Ω input terminals of an amplifier is 100 μV. If the signal-to-noise ratio at this point is 1000 calculate the input noise voltage.

2.26. An amplifier has a bandwidth of 1 MHz. What would be the effect on the signal-to-noise ratio at the output of the amplifier of (i) increasing the bandwidth to 2 MHz, and (ii) reducing the bandwidth to 750 kHz. State any assumptions made in part (ii).

2.27. A 90 MHz carrier is frequency-modulated by a 10 kHz sinusoidal signal. List the component frequencies which may be present in the modulated wave.

2.28. A 2 kHz signal frequency modulates a carrier with a modulation index of (i) 5 and (ii) 8. What is the frequency deviation of the carrier in each case?

2.29. A frequency-modulated wave has a modulation index of 6 and the frequency deviation is 12 kHz. Calculate the modulating frequency.

3 Carrier Frequencies, Bandwidths, and Maximum Power Transfer

A number of different methods of modulation are employed in telecommunication engineering to enable the best use to be made of the frequency spectrum available in a given transmission medium. For each channel in a communication system a *carrier frequency* must be chosen to position the channel in the required part of the frequency spectrum. To obtain the maximum utilization of the transmission medium and to minimize costs, the channel bandwidth must be as narrow as possible; it must, however, be wide enough to pass all the significant frequency components of the signal. The signals to be transmitted over the various types of communication system fall into one of five classes: (*a*) telephony, (*b*) telegraphy, (*c*) music, (*d*) television and (*e*) data transmission. This chapter will discuss the various factors that must be considered when determining the carrier frequencies and bandwidths for a particular system.

Line Communication Systems

(1) Telephony

Telephone cables are capable of transmitting a band of frequencies well in excess of the normal speech frequency range and can therefore be used to carry single-sideband amplitude-modulated frequency-division multiplex telephony systems. A number of channels are made available by such a system, each channel being allocated a different carrier frequency. The range of carrier frequencies depends upon the number of channels provided by the system and the part of the frequency spectrum which the system is to occupy.

The number of channels that can be carried by a single cable pair is primarily determined by the attenuation of the cable at the highest frequency to be transmitted. The greater the cable attenuation the closer together must the line amplifiers be spaced in order to prevent the signal level falling below a predetermined value. The line amplifiers are housed in repeater station buildings and these are expensive (some modern transistorized coaxial systems utilize manholes). Thus the signals of a multi-channel telephony system should be transmitted in the lowest frequency band possible.

Most of the multi-channel telephony systems in use in the United Kingdom consist of a suitable combination of CCITT† 12-channel carrier groups. By international agreement each channel in such a group must provide an audio bandwidth of 300–3400 Hz, which is sufficient to provide "commercial-quality" speech. The CCITT 12-channel group will be considered in some detail in Chapter 5, but its basic principle is shown in Fig. 3.1. Each channel has a different carrier fre-

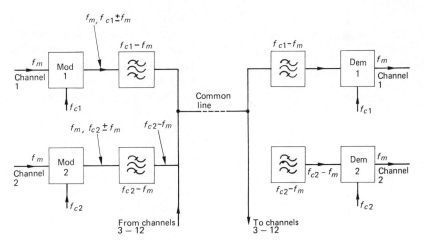

Fig. 3.1 Principle of a 12-channel telephony system

quency that is amplitude modulated by the audio-frequency signal applied to the channel. The lower sidebands of each channel are selected by the channel filters and the composite signal is transmitted over the common line. At the receiving end, channel filters select the frequencies proper to each channel, and demodulators extract the original audio intelligence.

To keep inter-channel crosstalk to an acceptably low level, each filter is required to attenuate signals more than 600 Hz outside the 3.1 kHz channel bandwidth by at least 70 dB. To achieve this discrimination with a spacing of only 900 Hz

† C.C.I.T.T.: International Telegraph and Telephone Consultative Committee.

50 CARRIER FREQUENCIES, BANDWIDTHS, AND MAXIMUM POWER TRANSFER

Table 3.1

Channel no.	Carrier frequency (kHz)	Channel filter passband (kHz)
1	108	104.6–107.7
2	104	100.6–103.7
3	100	96.6– 99.7
4	96	92.6– 95.7
5	92	88.6– 91.7
6	88	84.6– 87.7
7	84	80.6– 83.7
8	80	76.6– 79.7
9	76	72.6– 75.7
10	72	68.6– 71.7
11	68	64.6– 67.7
12	64	60.6– 63.7

between channels, *crystal filters* must be used. Crystal filters have very sharp attenuation/frequency characteristics, but are economic only at frequencies greater than about 60 kHz. For this reason the channel carrier frequencies start at 64 kHz for channel 12 and increase in 4 kHz steps to 108 kHz for channel 1. The lower sidebands of each channel are selected and so the bandwidth of the complete 12-channel system is 60.6–107.7 kHz (60.6 kHz = 64 − 3.4 kHz and 107.7 kHz = 108 − 0.3 kHz).

The channel carrier frequencies are specified by the CCITT and listed in Table 3.1. The table also gives details of the passband of each channel filter; it should be noted that the bandwidths correspond to an audio bandwidth of 300–3400 Hz.

The transmitted bandwidth is therefore 60.6–107.7 kHz, or approximately 60–108 kHz.

(2) Telegraphy

Many teleprinter circuits are routed wholly or partly over a multi-channel voice-frequency (MCVF) telegraphy system. The basic principle of a MCVF telegraphy system is similar to that of the multi-channel telephony system shown in Fig. 3.1. The teleprinter signal is applied to one channel of the system and modulates the channel carrier frequency. The output from a channel send equipment consists of a series of pulses of carrier as shown in Fig. 3.2. Such a waveform constitutes 100% double-sideband amplitude modulation.

The bandwidth required to transmit a teleprinter signal depends upon the characters sent, but the maximum bandwidth is demanded when alternate marks and spaces are

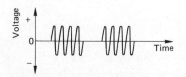

Fig. 3.2 Output waveform of a telegraph modulator

CARRIER FREQUENCIES, BANDWIDTHS, AND MAXIMUM POWER TRANSFER

transmitted, i.e. letters R and Y. The periodic time of one cycle of the waveforms for R and Y is 40 ms, and so the fundamental frequency of the waveform is 1000/40, or 25 Hz. A minimum channel bandwidth of 50 Hz is therefore required.

The choice of the carrier frequencies for the individual channels of a MCVF telegraphy system is determined primarily by the need to minimize inter-channel interference. The double-sideband outputs of the individual channels are combined and transmitted over the common path. Any non-linearity in the output/input characteristics of the line amplifiers, or other common equipment, will result in the presence of both *harmonic* and *intermodulation* products. For example, if f_1 and f_2 are two of the carrier frequencies, then the harmonic products are $2f_1$, $3f_1$, $4f_1$, etc., and $2f_2$, $3f_2$, $4f_2$, etc., and the intermodulation products are $f_1 \pm f_2$, $2f_1 \pm f_2$, $2f_2 \pm f_1$, etc. However, the products of greatest amplitude, and hence of the most importance, are $2f_1$, $2f_2$ and $f_1 \pm f_2$. To minimize inter-channel interference the channel carrier frequencies must be chosen to ensure that the important products do not fall within the passbands of the channel filters. This is achieved in practice by choosing odd harmonics of 60 Hz, starting with the seventh, as the channel carrier frequencies. Thus the channel carrier frequencies start at 420 Hz for channel 1 and increase in 120 Hz steps to 3180 Hz for channel 24.

FACSIMILE TELEGRAPHY is the transmission and reception of still pictures and diagrams. The picture to be transmitted is mounted on a circular drum which is threaded on to a lead-screw so that for each complete revolution it makes the drum move longitudinally a distance equal to the pitch of the screw thread. For each position of the drum an elemental area of the picture is illuminated by a spot of light, and as the drum revolves the entire picture is scanned by this light spot. The transmitting equipment is arranged to produce a voltage output proportional to the intensity of the light reflected from each elemental area of the picture, and thus the output voltage fluctuates in accordance with the nature of the picture.

The output waveform contains frequencies from zero, when the picture is of uniform shade, to a maximum frequency that is a function of the nature of the picture and the speed of scanning.

The picture is made up of a number of elemental areas which for equal horizontal and vertical definition should be square in shape. The maximum frequency of the output waveform will occur when the picture is of a chessboard pattern as in Fig. 3.3.

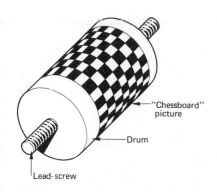

Fig. 3.3 Picture demanding maximum bandwidth from a facsimile telegraphy system

Let d be the diameter of the drum, n the number of revolutions per second made by the drum and l the width of each scanning line, i.e. the pitch of the screw thread.

The length of one line is equal to the circumference of the drum, i.e. πd, and since the elemental areas are square in shape, the number of picture elements per line is $\pi d/l$. The number of elements scanned per second, N, is the number of picture elements per line times the number of drum revolutions per second. Therefore

$$N = \frac{\pi d n}{l} \quad \text{elements/second} \tag{3.1}$$

Now, each pair of black and white elements will produce one cycle of the output waveform and so the maximum frequency that must be transmitted is

$$f = \frac{\pi d n}{2l} \quad \text{hertz} \tag{3.2}$$

Typically, f is of the order of 550 Hz, so that a bandwidth of about 0–550 Hz is required. Very often the pictures are to be transmitted between points of wide geographical separation and then the signals will require amplification at intervals along the line. It is difficult to amplify d.c. signals and it is therefore usual for the picture signal to amplitude-modulate a carrier before transmission so that a more suitable part of the frequency spectrum of the line is occupied. The CCITT recommend two carrier frequencies: 1300 Hz for transmission on audio-frequency amplified circuits, and 1900 Hz for circuits to be routed over a multi-channel telephony system. The bandwidth required to transmit the amplitude-modulated picture signal is

$$\text{Bandwidth} = \frac{\pi d n}{l} \quad \text{hertz} \tag{3.3}$$

EXAMPLE 3.1

A facsimile telegraphy system employs a drum of diameter 6.6 mm that makes one revolution per second and has $5\frac{1}{3}$ scanning lines per millimetre. The output waveform of the system modulates a 1900 Hz carrier; calculate the required bandwidth.

Solution
From eqn. (3.3),

$$\text{Bandwidth} = \frac{\pi \times 6.6 \times 10^{-3} \times 1}{\frac{1}{5\frac{1}{3}} \times 10^{-3}} = 111 \text{ Hz} \quad (Ans.)$$

CARRIER FREQUENCIES, BANDWIDTHS, AND MAXIMUM POWER TRANSFER

Radiocommunication Systems

(1) General

The radiocommunication systems in use today may be divided into one of two main classes: firstly, *broadcasting*—both radio and television—and, secondly, *radio links* for providing point-to-point telephonic communication.

Table 3.2 gives the classification of the various frequency bands.

Table 3.2

Frequency band	Classification	Abbreviation
10–30 kHz	very-low frequencies	v.l.f.
30–300 kHz	low frequencies	l.f.
300–3000 kHz	medium frequencies	m.f.
3–30 MHz	high frequencies	h.f.
30–300 MHz	very-high frequencies	v.h.f.
300–3000 MHz	ultra-high frequencies	u.h.f.
3–30 GHz	super-high frequencies	s.h.f.
30–300 GHz	extra-high frequencies	e.h.f.

(2) Sound Broadcasting

A sound broadcasting system is one in which programmes of mixed entertainment, news and educational content are made available to a large number of people with radio receivers. At the lower frequencies a single transmitter can serve the whole country, but at higher frequencies the range of a particular station is much smaller and it becomes necessary to have a number of transmitting stations located in different parts of the country.

The BBC broadcast their Radio 1, 2, 3 and 4 programmes at a number of different frequencies in the medium and v.h.f. bands and at one frequency in the low-frequency band. In the medium-frequency band, amplitude-modulated d.s.b. transmissions are used with carrier frequencies in the range 647–1546 kHz and having a bandwidth of approximately 9000 Hz. By international agreement, in Europe carrier frequencies in the medium-frequency band are spaced at 9000 Hz intervals and, consequently, the highest audio frequency transmitted is in the region of 4500 Hz, i.e. a bandwidth of 9000 Hz.

The use of a 4.5 kHz audio bandwidth for medium-frequency broadcast transmissions means poor reproduction of music because many of the higher harmonics produced by musical instruments are suppressed. For reasonably high-quality reception of music, an audio bandwidth of at least 15 kHz is required. This occupies an r.f. bandwidth of 30 kHz, which cannot be accommodated in the congested medium-frequency band. High-quality broadcast transmissions are therefore provided in the v.h.f. band using carrier frequencies in the range of 88.1–96.8 MHz. V.H.F. signals have a limited range of a hundred kilometres or so and so a number of stations are required within a fairly small area. The carrier frequencies allocated to the stations in a given area are about 200 kHz apart, to minimize inter-station interference. The wide carrier-frequency spacing allows frequency modulation to be used; this gives a better signal/noise ratio than amplitude modulation but requires a much wider bandwidth (in the BBC v.h.f. broadcast system an r.f. bandwidth of 180 kHz is necessary to provide the 15 kHz audio bandwidth).

Sound broadcast transmissions are also radiated in certain parts of the high-frequency band:

5.950–6.20 MHz, 7.1–7.3 MHz, 9.5–9.775 MHz,
11.7–11.975 MHz, 15.1–15.45 MHz, 17.7–17.9 MHz,
21.45–21.75 MHz and 25.6–26.1 MHz,

with a bandwidth of 10 kHz. Transmissions in these frequency bands are employed in Europe for international broadcast programmes as, for example, BBC programmes to Europe and to North Africa.

(3) Television Broadcasting

Three television services are available to the British viewing public; two are provided by the BBC and the third by the various commercial television companies but transmitted by the Independent Broadcasting Authority (IBA). Colour and monochrome television signals are transmitted in the u.h.f. band by both the BBC (BBC 1 and BBC 2) and the IBA. These programmes use vestigial sideband amplitude modulation for the picture signal, and frequency modulation for the sound signal. The vision carrier frequencies used are in the bands 471.25–575.25 and 615–847.25 MHz, providing a video bandwidth of 5.5 MHz. The sound carrier frequencies are spaced 6.0 MHz above the associated vision carrier.

(4) Point-to-Point

Post Office high-frequency radio links are used to route telephone calls to overseas destinations where line communication links via submarine cable or earth satellites are not available or have inadequate capacity. Different carrier frequencies are employed for each direction of transmission; and receiving stations are geographically situated well apart to minimize interference between signals passing in different directions. Radio links may be used to carry ordinary telephone conversations or to carry facsimile telegraphy or Press broadcasts. The majority of radio links have carrier frequencies in the high-frequency band and generally use a form of s.s.b. amplitude modulation. Radio-telegraphy services, such as the Press Broadcast Service, which is provided for the use of news agencies, are also generally provided in the high-frequency band.

A large number of frequency bands have been allocated to h.f. radio links; some of these, for example, are

3.5–3.9 MHz, 5.73–5.95 MHz, 9.04–9.5 MHz,
13.36–14 MHz, 21.75–21.87 MHz, 26.1–27.5 MHz.

The bandwidth per channel is 2.5–3 kHz.

(5) Mobile Systems

Telephonic and telegraphic services to ships at sea are provided at a number of specified frequencies in the medium-, high-, and very high-frequency bands. Short-range telephony services, up to about 80 km in distance, are operated at frequencies in the v.h.f. band, 156–163 MHz. Telephony services to ships at distances in excess of 80 km and less than about 1000 km are provided on specified carrier frequencies in the band 1.6–3.8 MHz. Ship-to-shore telegraphy services are operated, using the Morse code, in the band 405–525 kHz, except for some ships in northern seas which may use specified channels in the 1.6–3.8 MHz band. Longer distance services to ships at sea are operated at the following frequencies in the h.f. band; 4, 6, 8, 13, 17 and 22 MHz. 500 kHz (telegraphy) and 2182 kHz (telephony) are used as international distress and calling signals. A mobile radio telephone service which provides telephonic communication between mobile radio telephones in motor cars and the public telephone network exists in several parts of the country. The systems operate on frequency modulation and carrier frequencies in the band 158–164.4 MHz, with 25 kHz bandwidth.

Many organizations, e.g. police, taxi-cab firms, and ambulances, use base-to-mobile systems to provide telephonic communication between a headquarters and several mobile units. Systems of this kind are operated in the v.h.f. band, and in the u.h.f. band.

A considerable number of different frequency bands, of various widths, have been allocated to land, sea and air mobile services in the v.h.f. and u.h.f. bands. The complete list of

Table 3.3

Frequency band	Used by	Channel bandwidth (kHz)
71.5– 78.0 MHz	private land	12.5
80.0– 85.0	police and fire	12.5
85.0– 88.0	private land	12.5
97.0–102.0	police and fire	12.5
105.0–108.0	private land	12.5
108.0–136.0	aero	25
138.0–141.0	private land	12.5
146.0–148.0	police and fire	12.5
156.0–163.0	maritime	25
165.0–173.0	private land	12.5
425.0–449.5	private land	12.5
451.0–452.0	police and fire	25
453.0–462.5	private land	25
465.0–466.0	police and fire	25

frequencies is too long to include in this book and so Table 3.3 gives some examples.

(5) Multi-Channel Telephony/Television Links

In the u.h.f. and s.h.f. bands, multi-channel f.d.m. telephony systems, known as radio-relay systems, are an alternative to line systems for the provision of large numbers of circuits between two points. In Great Britain microwave point-to-point links are in the following bands:

1.7–1.9 GHz (the 2 GHz spur band)
1.9–2.3 GHz (the 2 GHz main band)
3.7–4.2 GHz (the 4 GHz band)
5.85–6.425 GHz (the lower 6 GHz band)
6.425–7.11 GHz (the upper 6 GHz band)
10.7–11.7 GHz (the 11 GHz band).

The 2 GHz spur and main bands, the upper 6 GHz and the 11 GHz bands are split into a number of channels each of which is capable of handling 960 telephony channels or a television signal. The 2 GHz spur band is used to provide spur links within a city and for low-capacity links in remote areas. All other bands, including the 4 GHz and lower 6 GHz bands which are each divided into channels capable of accommodating 1800 channel telephony systems, are used for inter-city links, or links between a television studio and a television transmitter.

Many international telephony circuits are routed over multi-channel systems which are themselves routed over a satellite

CARRIER FREQUENCIES, BANDWIDTHS, AND MAXIMUM POWER TRANSFER

communication link. The internationally agreed frequency bands for satellite systems are

5.925–6.425 GHz for transmission from the Earth to the satellite

3.7–4.2 GHz for transmission from the satellite to the Earth.

Maximum Power Transfer

Many instances occur in electronic and telecommunication engineering where a source of e.m.f. is to be connected to a load and maximum power is to be transferred to the load from the source. Consider the simple circuit shown in Fig. 3.4a, in which a source of e.m.f. 10 V and resistance 10 Ω is connected across a variable resistance R. The current flowing in the circuit is given by $I = 10/(10+R)$ and the power P dissipated in the load is

$$P = I^2 R = [10/(10+R)]^2 R$$

If the value of R is increased in a number of steps starting from 0 Ω it is found that the power dissipated in R increases at first and thereafter it decreases. Suppose that R is increased in 2 Ω steps from 0 Ω to 16 Ω. The values then obtained for both the current I flowing in the circuit and the power P dissipated in the load R are given in Table 3.4 and are shown plotted in Fig. 3.4b.

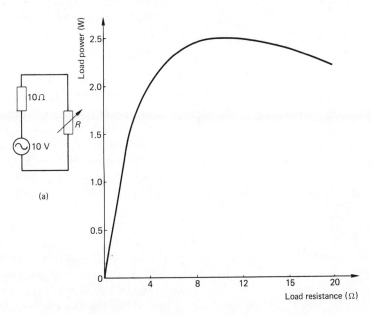

Fig. 3.4 Maximum power transfer

Table 3.4

R	0	2	4	6	8	10	12	14	16	18	20
I	1	0.833	0.714	0.625	0.556	0.5	0.455	0.417	0.385	0.357	0.333
P	0	1.389	2.041	2.344	2.469	2.5	2.479	2.431	2.367	2.296	2.222

It is clear that the maximum power is dissipated in the load resistance when it has the same value as the source resistance. Hence, for maximum power to be transferred from a source to a load, the resistance of that load must be equal to the resistance of the source.

EXAMPLE 3.2

A source of e.m.f. 2 V and internal resistance 5 Ω is connected to a resistive load. Determine (i) the resistance of the load for maximum power to be dissipated in it and (ii) the value of this maximum power.

Solution

(i) For maximum power transfer $R_L = R_S$. Therefore
$$R_L = 5\,\Omega \quad (Ans.)$$

(ii) With $R_L = 5\,\Omega$ the current flowing in the circuit is $2/(5+5) = 0.2$ A and the power dissipated in the load is
$$P_{max} = 0.2^2 \times 5 = 200\,\text{mW} \quad (Ans.)$$

Very often the load resistance does not have the same value as the source resistance and if maximum power transfer is to be achieved a transformer must be used to transform the load resistance to the required value.

The Use of a Transformer for Resistance Matching

In a transformer having a magnetic core almost all the flux set up by the current in the primary winding links with the turns of the secondary winding. For such a transformer, the ratio of the voltage appearing across the terminals of the secondary winding to the voltage applied across the terminals of the primary winding is very nearly equal to the ratio of the number of secondary turns to the number of primary turns. Thus, referring to Fig. 3.5,

$$\frac{V_s}{V_p} = \frac{N_s}{N_p} \quad (3.4)$$

where V_s, V_p are respectively the voltages across the secondary and primary windings and N_s, N_p are respectively the number of turns in the secondary and primary windings.

Also, the current transformation ratio is the inverse of the voltage transformation ratio, i.e.

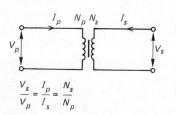

Fig. 3.5 Representation of a transformer

CARRIER FREQUENCIES, BANDWIDTHS, AND MAXIMUM POWER TRANSFER

$$\frac{I_s}{I_p} = \frac{N_p}{N_s} \quad (3.5)$$

Now $I_p = V_p/R_p$ and $I_s = V_s/R_s$ where R_p and R_s are the resistances of the primary and secondary circuits respectively; hence, substituting in equation (3.5),

$$\frac{V_s}{R_s} \Big/ \frac{V_p}{R_p} = \frac{N_p}{N_s}$$

$$\frac{V_s R_p}{V_p R_s} = \frac{N_p}{N_s}$$

But, from equation (3.4),

$$\frac{V_s}{V_p} = \frac{N_s}{N_p}$$

Therefore

$$\frac{R_p}{R_s} = \left(\frac{N_p}{N_s}\right)^2$$

$$\frac{N_p}{N_s} = \sqrt{\left(\frac{R_p}{R_s}\right)} \quad (3.6)$$

This relationship can be used to obtain a desired impedance transformation by suitable choice of the numbers of primary and secondary turns.

EXAMPLE 3.3

Determine the turns ratio of a transformer which is to be employed to match a 50 Ω load to a 1000 Ω source. (The term "match" means to make the impedance effectively connected across the source equal to the source impedance, see Fig. 3.6.)

Solution

$$\frac{N_p}{N_s} = \sqrt{\left(\frac{1000}{50}\right)} = \sqrt{20} = 4.47$$

Thus the primary winding must have 4.47 times as many turns as the secondary winding.

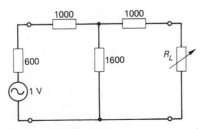

Fig. 3.6

60 CARRIER FREQUENCIES, BANDWIDTHS, AND MAXIMUM POWER TRANSFER

Exercises

3.1. State the frequency band in which v.h.f. frequency-modulated sound broadcast stations operate. The BBC system employs a rated system deviation of 75 kHz and accommodates a maximum modulating frequency of 15 kHz. Determine the r.f. bandwidth required. The sound signal of u.h.f. television employs frequency modulation with a rated system deviation of 50 kHz and a maximum modulating frequency of 15 kHz. Calculate the necessary r.f. bandwidth.

3.2. State the carrier frequencies and r.f. bandwidths required for each of the following radio systems:
  (a) 625 line colour television,
  (b) a radio-telephone system for telephone calls between subscribers in different countries,
  (c) an overseas radio-telegraphy call between a shipping company and one of its ships,
  (d) public ship-to-shore long-distance telephony.
Explain what is meant by single-sideband operation of a radio system.

3.3. Briefly explain why a carrier is used in radio and line circuits. State the carrier frequencies and bandwidths suitable for the following applications:
  (a) a local broadcast service,
  (b) a long-distance overseas broadcast service,
  (c) a private mobile radio-telephone service to taxis,
  (d) a long-distance overseas point-to-point radio-telegraphy service to ships,
  (e) a public ship-to-shore telephone service. (C&G)

3.4. (a) Discuss the reasons for the use of a carrier in radio and line transmission.
  (b) Suggest suitable values of carrier frequency and bandwidth for the following applications: (i) a long-distance overseas point-to-point single-sideband radio-telephony service, (ii) the provision of 24 voice-frequency telegraph channels over an audio-frequency cable pair, (iii) a sound broadcast service to serve a relatively small area, (iv) a private mobile radio-telephone service. (C&G)

3.5. (a) Briefly explain why the frequency band for commercial quality speech is generally restricted to the range 300–3400 Hz.
  (b) State typical radio bandwidths for the following applications: (i) a national broadcasting service in the medium waveband, (ii) a private mobile radio service in the v.h.f. band, (iii) a single-sideband overseas radio-telephony service in the h.f. band, (iv) a ship-to-shore teleprinter service in the h.f. band.
  (c) Sketch, on common axes, the following waveforms: (i) a sinusoidal tone of amplitude 2.5 V, (ii) the third harmonic of this tone with an amplitude of 1.5 V, (iii) the composite waveform formed by the waveforms in (i) and (ii) above.
(C&G)

3.6. Explain the principle of operation of a facsimile telegraphy system. For what purposes are such systems employed?
A facsimile telegraphy system employs a drum of 7 mm diameter that revolves 1.5 times per second. If the drum has 5 scanning lines per mm, determine the maximum frequency contained in the transmitted waveform. If the picture signal is used to amplitude-modulate a 1300 Hz carrier calculate the required bandwidth.

CARRIER FREQUENCIES, BANDWIDTHS, AND MAXIMUM POWER TRANSFER

3.7. State the maximum power transfer theorem for a resistive source and load.

A source of e.m.f. 12 V and impedance 600 Ω is required to deliver maximum power to a 1000 Ω load resistance. Determine (i) the turns ratio of a transformer that could be used to connect the load to the source, and (ii) the maximum power then dissipated in the load resistance.

3.8. State the maximum power transfer theorem. A source of 600 Ω impedance has its e.m.f. adjusted to be 1 V and is then connected to variable load R_L via the network in Fig. 3.6. Determine (i) the value of R_L for it to dissipate the maximum power, and (ii) the value of this maximum power.

3.9. State the maximum power transfer theorem. A source of e.m.f. 50 V and internal impedance 600 Ω is connected across a variable resistance R. Plot graphs to show how the powers dissipated in (i) the load resistance R, and (ii) the source resistance vary as R is adjusted from 100 Ω to 1000 Ω in 100 Ω steps. From your graphs state (iii) whether the maximum power transfer theorem has been confirmed, and (iv) whether the power dissipated in the source resistance also exhibits a maximum value.

Short Exercises

3.10. List the frequency bands in common use for radio and line systems.

3.11. List the channel carrier frequencies and channel bandwidths for the CCITT 12-channel carrier group. What do the initials CCITT stand for?

3.12. A source of e.m.f. 2 V and internal resistance 5 Ω is connected to a variable resistance R. Calculate and plot a graph of the power dissipated in R as R is varied from 1 Ω to 10 Ω.

3.13. (a) State the function of a carrier.

(b) State the relationship between the baseband signal and the carrier frequency.

(c) State the carrier frequencies, audio bandwidths and radio bandwidths used for the broadcasting of speech and music in (i) the medium waveband, (ii) the h.f. waveband and (iii) the v.h.f. band.

3.14. Determine the turns ratios of transformers used to convert (i) 8 Ω, (ii) 12 Ω, (iii) 16 Ω into a 1000 Ω resistance.

3.15. List the frequency bands and channel bandwidths that have been allocated for mobile use by (i) police and fire services and (ii) taxicabs.

3.16. What audio bandwidths are transmitted by (i) medium-wave sound broadcasts, (ii) v.h.f. sound broadcasts and (iii) the sound section of u.h.f. television broadcasts? What r.f. bandwidth is required in each case?

4 Filters

In both radio and line systems the need often arises for a group of frequencies contained within a wider frequency band to be transmitted while all other frequencies are suppressed. A *filter* is a circuit which has the ability to discriminate between signals at different frequencies because it has an attenuation that varies with frequency in a particular manner. If a signal containing components at a number of different frequencies is applied to the input of a filter, only some of those components will appear at its output terminals, the remainder having been greatly attenuated and so effectively suppressed.

Four basic types of filter are available for use in telecommunication systems: the low-pass, the high-pass, the band-pass and the band-stop. Fig. 4.1 shows the circuit symbols for each of these filters. Filters can be designed using one of the following different techniques: inductor-capacitor filter, crystal filters and active filters.

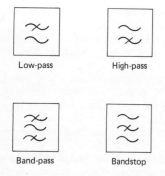

Fig. 4.1 Types of filter network

Inductor-Capacitor Filters

The transmission of an unwanted frequency through a network can be prevented either by connecting a high impedance (at that frequency) in series, and/or by connecting a low impedance in shunt, with the signal path. The high series impedance will oppose the flow of currents, at the unwanted frequencies, through the network, and the shunt impedance will bypass unwanted currents to earth. The necessary high and low values of impedance can be obtained by the use of inductors and capacitors of appropriate value. This is because an inductor has a reactance which is directly proportional to frequency and the reactance of a capacitor is inversely proportional to frequency. If a particular band of frequencies is to be transmitted by a filter, series and shunt impedances are required which reach a maximum or a minimum value at the centre of the frequency band. Such impedances can be obtained by the use of series and parallel resonant circuits.

The Prototype or Constant-k Filter

A LOW-PASS filter should be able to pass, with zero attenuation, all frequencies from zero up to a certain frequency which is known as the CUT-OFF FREQUENCY f_c. At frequencies greater than the cut-off frequency the attenuation of the filter will increase with increase in frequency up to a very high value. The basic prototype (or constant-k) T and π low-pass filter circuits are shown in Fig. 4.2a and b respectively. For both circuits the total series impedance is ωL and the total shunt impedance is $1/\omega C$. The term "constant-k" is used to denote that the product of the series and shunt impedances is a constant at all frequencies.

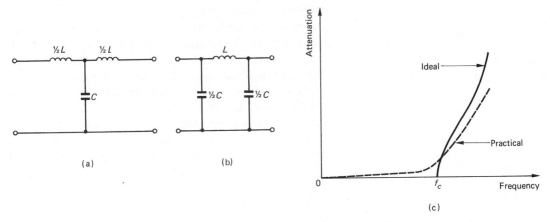

Fig. 4.2 The constant-k low-pass filter

Suppose a voltage source of variable frequency is applied across the input terminals of the filter. At low frequencies the reactance of the series inductor L is low and the reactance of the shunt capacitor C is high; at these frequencies therefore the inductance offers little opposition to the flow of current while the capacitance has zero shunting effect. Low-frequency signals are propagated through the filter without loss. As the frequency of the input signal is increased, the inductive reactance will fall until, at the cut-off frequency f_c, the attenuation of the filter suddenly increases. Thereafter, the attenuation of the filter rises rapidly with increase in frequency. The ideal attenuation/frequency characteristic of a constant-k low-pass filter is shown by Fig. 4.2c. In practice, an inductor inevitably possesses some resistance and because of this the filter does introduce some attenuation into the passband; also, the attenuation does not rise so sharply at the cut-off frequency. The practical attenuation/frequency characteristic of a low-pass filter is shown by the dotted line of Fig. 4.2c.

The action of a HIGH-PASS filter is to transmit all frequencies which are higher than its cut-off frequency and to prevent the passage of all lower frequencies. Figs. 4.3a and b give the circuits of T and π constant-k high-pass filters. At low frequencies the series capacitance C has a high reactance and the shunt inductive reactance is low, so low-frequency signals are attenuated as they travel through the filter. At high frequencies, on the other hand, the series reactance is low and the shunt reactance is high and the filter offers zero attenuation. The attenuation/frequency characteristic of the ideal high-pass filter is shown in Fig. 4.3c, while the dotted curve shows how the presence of resistance modifies the ideal characteristic.

Fig. 4.4a shows the circuit of a T constant-k BAND-PASS filter and Fig. 4.4b shows its ideal and practical attenuation/frequency characteristics. Ideally, the filter passes, with zero attenuation, a particular band of frequencies and offers considerable attenuation to all frequencies outside of this passband. The required characteristic is obtained by using two series-tuned circuits as the series impedance and a single parallel-tuned circuit as the shunt impedance. The three circuits are arranged to be resonant at the same frequency. For signals at or near this common resonant frequency, the series reactance is low and the shunt reactance is high so that the filter offers, ideally, zero attenuation. At frequencies either side of the required passband the tuned circuit's impedances have varied to such an extent that considerable attenuation is offered.

The fourth kind of filter, which is very much less often used, is the BAND-STOP filter. This type of filter, shown in Fig. 4.5a, provides a large attenuation to signals whose frequencies are within a particular frequency band. The ideal and the practical attenuation/frequency characteristics of a band-stop filter are shown in Fig. 4.5b.

m-Derived Filters

The constant-k filter suffers from two main disadvantages; these are (a) the attenuation/frequency characteristic does not rise at the cut-frequency as sharply as is often required, and (b) its input and output impedances vary with frequency. In an *m*-derived filter, components are added to the basic constant-k circuit to ensure that a very high attenuation is obtained at a particular frequency. In this book only the low-pass *m*-derived filter will be discussed.

Consider the T constant-k low-pass filter shown in Fig. 4.2a. The attenuation can be made to reach a very high value at a

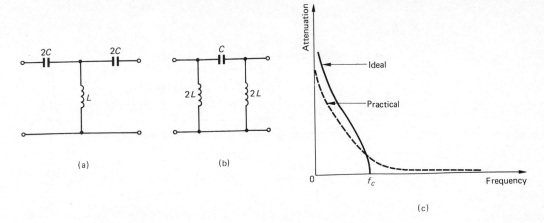

Fig. 4.3 The constant-k high-pass filter

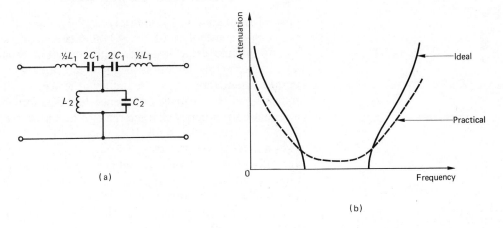

Fig. 4.4 The constant-k band-pass filter

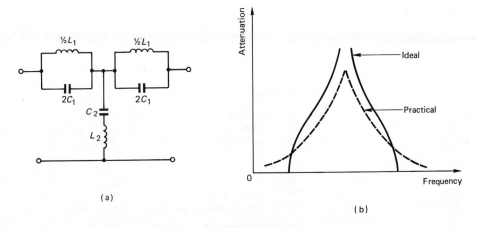

Fig. 4.5 The constant-k band-stop filter

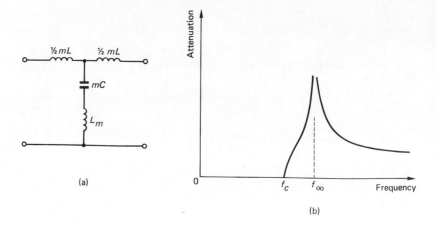

Fig. 4.6 The *m*-derived T low-pass filter

desired frequency f_∞ by multiplying the values of its components by a constant m ($0 < m < 1$) and connecting an inductor L_m of suitable value in series with the shunt capacitor (Fig. 4.6a). Neglecting circuit resistance, the filter will now have infinite attenuation at the frequency at which the shunt series-tuned circuit is resonant and hence has zero impedance. The attenuation/frequency characteristic of the *m*-derived low-pass filter is shown by Fig. 4.6b; it can be seen that the attenuation rises above zero at the cut-off frequency f_c and increases to a very high value at some frequency f_∞. Unfortunately, at frequencies above f_∞ the attenuation of the filter falls and will eventually reach a low value.

An *m*-derived π filter can be constructed in a similar manner; the series inductance L and the shunt capacitance C are both multiplied by the constant-*m*, and a capacitor C_m is connected in parallel with the modified series inductor (Fig. 4.7). The frequency at which the maximum attenuation takes place depends upon the value of m that is chosen. Fig. 4.8 shows the relationship between m and the frequency of maximum attenuation.

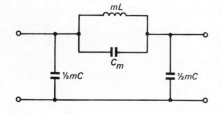

Fig. 4.7 The *m*-derived π low-pass filter

EXAMPLE 4-1

A constant-k T low-pass filter has cut-off frequency of 12 kHz. Determine the value of m for the corresponding *m*-derived filter in order to position maximum attenuation at (i) 13.8 kHz and (ii) 19.8 kHz.

Solution
(i) $13.8/12 = 1.15$
Therefore, from the graph, $m = 0.5$ (*Ans.*)

(ii) $19.8/12 = 1.65$
Therefore, from the graph, $m = 0.8$ (*Ans.*)

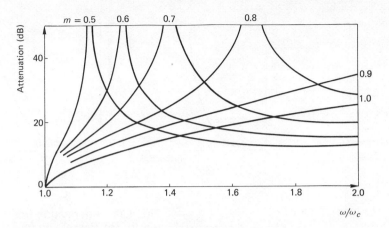

Fig. 4.8 Showing the relationship between m and the frequency of maximum attenuation

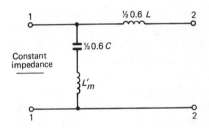

Fig. 4.9 m-derived half-section with $m = 0.6$

Fig. 4.9 shows the circuit of a half-section m-derived T low-pass filter. If m is chosen to be 0.6 the impedance measured at the terminals 1-1 does not vary with frequency, while the impedance "seen" looking into terminals 2-2 varies with frequency in the same way as the impedance of a whole section, constant-k or m-derived. Such a half-section can be used to terminate both ends of a filter to ensure that the filter presents constant values of impedance at both its input and its output terminals.

The attenuation of an m-derived filter section decreases with increase in frequency above the frequency of maximum attenuation. This disadvantage can be overcome by connecting a constant-k filter section in cascade to ensure substantial attenuation at all frequencies above cut-off. The block diagram of a composite filter is shown in Fig. 4.10a and its attenuation/frequency characteristic is given by Fig. 4.10b.

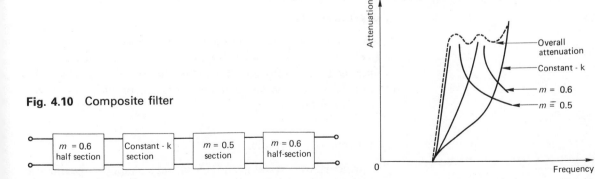

Fig. 4.10 Composite filter

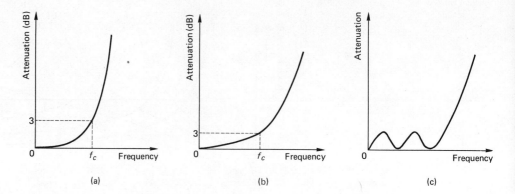

Fig. 4.11 Attenuation-frequency characteristics of (a) Butterworth, (b) Bessel and (c) Tchebyscheff low-pass filters

Modern Filter Designs

In recent years many L-C filters have been designed using different techniques to those used for the design of constant-k and m-derived filters. These methods allow a filter to be designed with a very accurate desired attenuation/frequency characteristic.

The Butterworth attenuation/frequency characteristic of a low-pass filter is shown in Fig. 4.11a; it is maximally flat in the passband and has an attenuation of 3 dB at the cut-off frequency f_c. The Bessel filter (Fig. 4.11b) introduces a constant time delay to all frequencies in the passband and has 3 dB loss at the cut-off frequency. It is evident that the attenuation of the Bessel filter does not rise as rapidly as that of the Butterworth filter. A more rapid increase in attenuation outside of the passband can be obtained by the Tchebyscheff filter (Fig. 4.11c) at the expense, however, of ripple in the passband. The relative merits of the three approaches to filter design are the same when applied to the other kinds of filters, e.g. high-pass.

Crystal Filters

For some applications the maximum selectivity a bandpass L-C filter can attain is inadequate, and in such cases a crystal filter can be employed. A crystal filter is one in which the required series and shunt impedances are provided by *piezoelectric crystals*.

Piezoelectric Crystals

A piezoelectric crystal is a material, such as quartz, having the property that, if subjected to a mechanical stress, a potential difference is developed across it, and if the stress is reversed a p.d. of opposite polarity is developed. Conversely, the application of a potential difference to a piezoelectric crystal causes the crystal to be stressed in a direction depending on the polarity of the applied voltage.

In its natural state, quartz crystal is of hexagonal cross-section with pointed ends. If a small, thin plate is cut from a crystal the plate will have a particular natural frequency, and if an alternating voltage at its natural frequency is applied across it, the plate will vibrate vigorously. The natural frequency of a crystal plate depends upon its dimensions, the mode of vibration and its original position or *cut* in the crystal. The important characteristics of a particular cut are its natural frequency and its temperature coefficient; one cut, the *GT cut*, has a negligible temperature coefficient over a temperature range from 0°C to 100°C; another cut, the *AT cut*, has a temperature coefficient that varies from about +10 p.p.m./°C at 0°C to 0 p.p.m./°C at 40°C and about +20 p.p.m./°C at 90°C. Crystal plates are available with fundamental natural frequencies from 4 kHz up to about 10 MHz or so. For higher frequencies the required plate thickness is very small and the plate is fragile; however, a crystal can be operated at a harmonic of its fundamental frequency and such *overtone* operation raises the possible upper frequency to about 100 MHz.

The electrical equivalent circuit of a crystal is shown in Fig. 4.12. The inductance L represents the inertia of the mass of the crystal plate when it is vibrating; the capacitance C_1 represents the reciprocal of the stiffness of the crystal plate; and the resistance R represents the frictional losses of the vibrating plate. The capacitance C_2 is the actual capacitance of the crystal (a piezoelectric crystal is an electrical insulator and is mounted between two conducting plates).

A series-parallel circuit, such as the one shown in Fig. 4.12 has two resonant frequencies: the resonant frequency of the series arm R-L-C_1, and the parallel resonance produced by C_2 and the effective inductance of the series arm at a frequency above its (series) resonant frequency.

If a pair of similar crystals is connected in the series arms of a lattice network and another pair connected in the shunt arms, as shown in Fig. 4.13a, the network will possess a band-pass characteristic, provided that the series resonant frequency of one pair is equal to the parallel-resonant frequency of the other pair. If the bandwidth is not wide enough it can be increased by the connection of an inductor of suitable value in series with the crystal (see Fig. 4.13b). Crystal filters of the lattice type are widely employed in multi-channel telephony systems. Other, simpler versions of crystal filters are commercially available for use as band-pass filters in radio receivers.

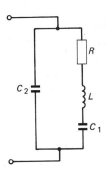

Fig. 4.12 Electrical equivalent circuit of a piezoelectric crystal

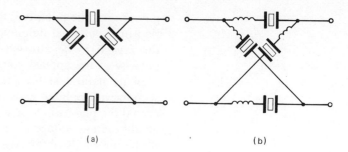

Fig. 4.13 Crystal filters   (a)   (b)

Filters in Parallel

Frequency-division multiplex is the transmission of two or more channels over a single circuit by the positioning of the channels at different parts of the frequency spectrum of that circuit. At the receiving end of an f.d.m. system the received signals must be directed to their correct channels and this is achieved by means of a number of band-pass filters connected in parallel. Since a small frequency gap, about 900 Hz, exists between adjacent passbands, the filters can be paralleled directly and connected to their common load, as shown in Fig. 4.14a. Each filter is then terminated by the load resistance in parallel with the output impedances of all the other filters. The two filters at the extreme ends of the system bandwidth only have another filter connected on one side of them. If these two filters are to be correctly terminated it is necessary to connect a *compensating network* in parallel with the load. This network provides the necessary impedance values at frequencies below the lowest frequency and above the highest frequency passed by the filters, in order for the two extreme filters to be correctly terminated.

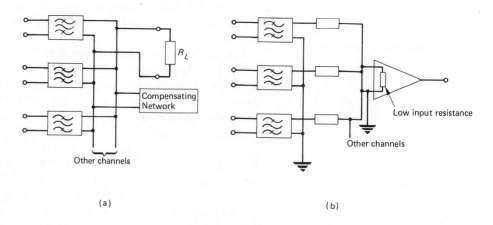

(a)   (b)

Fig. 4.14 Filters in parallel

In modern equipments a different method of paralleling band-pass filters is employed. The arrangement is illustrated by Fig. 4.14b. Each filter is terminated by an individual series resistor and the common, low value input resistance of an amplifier. The value of the series resistor is chosen so that each filter works into its correct load resistance.

Active Filters

Inductors are relatively large and bulky components particularly at the lower frequencies, and also possess core and winding losses that are difficult to predict accurately and which may vary with time, temperature and/or frequency. The need for an inductor in a filter network can be avoided if a resistor-capacitor network is used as the feedback network of an amplifier. A number of different types of active filter are possible but the kind most commonly used since integrated circuit *operational amplifiers* have become readily available is shown in Fig. 4.15. The circuit shown in Fig. 4.15a acts as a low-pass filter which can be given a Butterworth, a Bessel or a Tchebyscheff characteristic depending upon the values chosen for the various components. Changing over the positions of the resistors and the capacitors, as in Fig. 4.15b, produces a high-pass filter with the required type of attenuation/frequency characteristic. Lastly, a band-pass characteristic is obtained by connecting the resistance-capacitance network in the manner shown by Fig. 4.15c.

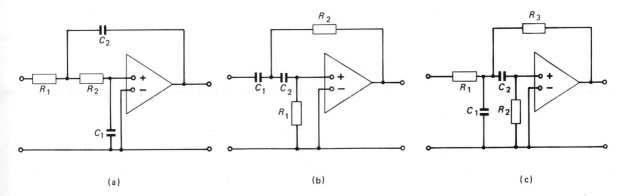

Fig. 4.15 Active filters: (a) low-pass, (b) high-pass and (c) band-pass

Exercises

4.1. What is meant by the term piezoelectric crystal? Draw a diagram to show how four such crystals can be connected to form a band-pass filter. Sketch the attenuation/frequency characteristic of such a filter.

4.2. Explain the electrical equivalent circuit of a quartz crystal and show how such a crystal can be used to advantage as an element in a band-pass filter. What are the advantages and disadvantages of crystal filters compared with coil-capacitor filters? (C&G)

4.3. Draw diagrams to show how (i) a constant-k low-pass filter and (ii) a T constant-k high-pass filter can be constructed. Explain the operation of each filter.

4.4. Draw the circuit diagram of a π constant-k bandpass filter and explain its operation.

4.5. Draw the circuit diagram of a π constant-k bandstop filter and explain its operation.

4.6. Draw the practical attenuation/frequency characteristic of a constant-k low-pass filter and use it to explain why the use of an m-derived filter is often necessary. Draw the circuit diagram of an m-derived T low-pass filter and explain how it operates.

4.7. Draw the practical attenuation/frequency characteristic of a constant-k high-pass filter. What are the disadvantages of this characteristic and how can they be overcome?

Draw the circuit diagram of an m-derived T high-pass filter.

4.8. Explain why filters are employed in communication equipment. What is meant by the terms *constant-k* and *m-derived*? Illustrate your answer by circuit diagrams and by sketches of attenuation/frequency characteristics for both low- and high-pass filters.

Show, with the aid of a block diagram, how filter sections could be connected to produce a low-pass filter having a cut-off frequency of 20 kHz with maximum attenuation at 23 kHz and constant input and output impedances.

4.9. (*a*) Draw the attenuation/frequency characteristics of the following types of high-pass filter: (i) constant-k, (ii) m-derived, (iii) Bessel, (iv) Butterworth and (v) Tchebyscheff.

(*b*) Why are crystal filters sometimes used?

Short Exercises

4.10. State the function of a filter. Draw attenuation/frequency characteristics for (i) a low-pass filter, (ii) a high-pass filter, (iii) a band-pass filter and (iv) a band-stop filter. Give also the symbol for each type of filter.

4.11. The band of frequencies 100 Hz to 50 kHz is applied in turn to the filters shown in Fig. 4.16. For each filter state the frequencies that appear at the output.

4.12. The band of frequencies 100 to 50 kHz is applied to the input of the cascaded filter network shown in Fig. 4.17. Determine the frequencies that appear at the output.

Fig. 4.16

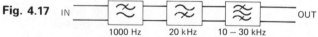

Fig. 4.17

4.13. What is an active filter? Why are active filters sometimes used in preference to L-C filters? What is an operational amplifier?

4.14. Draw and explain two ways in which a number of band-pass filters may be connected in parallel.

5 The Decibel ['desibel]

In telecommunication engineering an engineer is concerned with the transmission of intelligence from one point to another, the intelligence being transmitted in the form of electrical signals. A telecommunication system which carries such signals may consist of a number of links in tandem and, certainly, each link will consist of a number of different items, such as transmission lines and amplifiers, also connected in tandem. Each item will introduce a certain loss, or gain, of power into the system and the overall ratio (output power/input power), which is a measure of the efficiency of the system, is equal to the product of the power ratios of the individual items. Consider, as a simple example, the arrangement of Fig. 5.1.

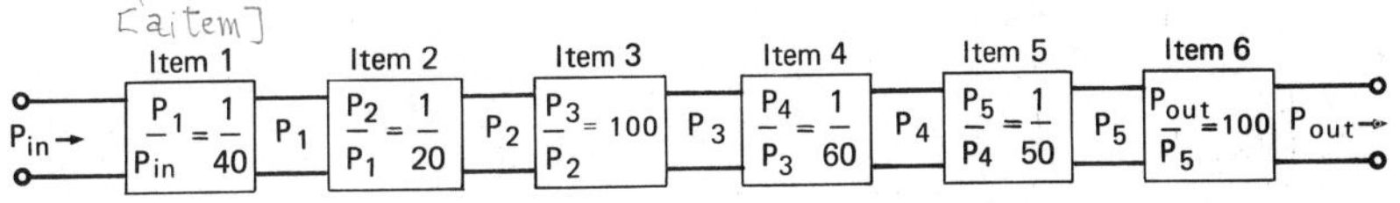

[aitem]

$P_{in} \rightarrow$ | Item 1 $\frac{P_1}{P_{in}}=\frac{1}{40}$ | P_1 | Item 2 $\frac{P_2}{P_1}=\frac{1}{20}$ | P_2 | Item 3 $\frac{P_3}{P_2}=100$ | P_3 | Item 4 $\frac{P_4}{P_3}=\frac{1}{60}$ | P_4 | Item 5 $\frac{P_5}{P_4}=\frac{1}{50}$ | P_5 | Item 6 $\frac{P_{out}}{P_5}=100$ | $\rightarrow P_{out}$

Fig. 5.1 The block schematic diagram of a generalized telecommunication system

If it is assumed that the input and output terminals of each item of equipment are matched,† then the input power to item 2 is equal to $P_{IN} \times \frac{1}{40}$, the input power to item 3 is equal to $P_{IN} \times \frac{1}{40} \times \frac{1}{20}$ and so on. Hence the input power P_{OUT} of the circuit is

$$P_{OUT} = P_{IN} \times \frac{1}{40} \times \frac{1}{20} \times 100 \times \frac{1}{60} \times \frac{1}{50} \times 100$$

or

$$\frac{P_{OUT}}{P_{IN}} = \frac{1}{240} = 0.004\,17$$

† Footnote is at the foot of p. 74.

74 THE DECIBEL

In this example simple figures have been selected and the arithmetic is easy. However, it should be appreciated that in practice the power ratios of the various items comprising a link would almost certainly involve less convenient numbers, and logarithmic tables or a slide rule or an electronic calculator would be needed. With logarithms the method of attack, again considering the simple example, would be to look up in tables the logarithm of each of the power ratios, add the values thus found, and then look up the antilogarithm of the result. Normally, logarithms to the base 10 are used.

$$\log_{10}\frac{P_{OUT}}{P_{IN}}$$

$$= \log_{10}\frac{1}{40} + \log_{10}\frac{1}{20} + \log_{10} 100$$

$$+ \log_{10}\frac{1}{60} + \log_{10}\frac{1}{50} + \log_{10} 100$$

$$= \log_{10} 1 - \log_{10} 40 + \log_{10} 1 - \log_{10} 20$$
$$+ \log_{10} 100 + \log_{10} 1 - \log_{10} 60$$
$$+ \log_{10} 1 - \log_{10} 50 + \log_{10} 100$$

$$\left(\text{since } \log\frac{a}{b} = \log a - \log b\right)$$

Now $\log_{10} 1 = 0$, therefore

$$\log_{10}\frac{P_{OUT}}{P_{IN}} = -\log_{10} 40 - \log_{10} 20 + \log_{10} 100$$
$$- \log_{10} 60 - \log_{10} 50 + \log_{10} 100$$
$$= -1.6021 - 1.3010 + 2 - 1.7782 - 1.6990 + 2$$
$$= -2.3803$$
$$= -3 + 0.6197$$

$$\frac{P_{OUT}}{P_{IN}} = \text{antilog}_{10}\,\bar{3}.6197 = 0.00417 \text{ as before.}$$

† The term "matched" means that the input terminals are closed with an impedance equal to the impedance seen looking into the terminals. For example, if the input impedance of item 1 is R ohms then the impedance of the source of the input power, P_{IN}, must also be R ohms. Similarly the output impedance of item 3 must be equal to the input impedance of item 4 and so on. If this is not so and a pair of terminals are "mismatched" some of the power arriving at those terminals is reflected back whence it came and the calculation of the overall loss, or gain, is much more difficult and beyond the scope of this book.

THE DECIBEL

Very often in practical calculations the power ratios involved are enormous and the direct use of power ratios would involve inconveniently large or small numbers. This suggests the possibility of directly quoting the power losses or gains of the items of equipment in a logarithmic form, so that the overall loss or gain, also quoted in logarithmic form, can be obtained by algebraic addition of the individual losses or gains. This, in fact, is the system used in practice, and the logarithmic unit is known as the decibel.

The Decibel

The decibel can be defined in the following way:

If the ratio of two powers P_1 and P_2 is to be expressed in decibels, the number of decibels, x, is given by

$$x = 10 \log_{10} \frac{P_1}{P_2} \qquad (5.1)$$

As an illustration consider again the system shown in Fig. 5.1.

Power ratio of item $1 = 10 \times -1.6021 = -16.021$ dB
$\simeq -16.02$ dB†
Power ratio of item $2 = 10 \times -1.3010 = -13.01$ dB
Power ratio of item $3 = 10 \times 2 = 20$ dB
Power ratio of item $4 = 10 \times -1.7782 = -17.78$ dB
Power ratio of item $5 = 10 \times -1.6990 = -17.0$ dB
Power ratio of item $6 = 10 \times 2 = 20$ dB

The overall power ratio is equal to the algebraic sum of these ratios, i.e. -23.81 dB. Therefore

$$10 \log_{10} \frac{P_{OUT}}{P_{IN}} = -23.81 \text{ dB}$$

The negative sign means that P_{OUT} is less than P_{IN}, i.e. there is a loss of 23.81 dB. (Note that to speak of a *loss* of -23.81 dB would mean a *gain* of 23.81 dB.)

EXAMPLE 5.1

Convert the following power ratios into decibels.
(a) $P_1/P_2 = 2$; (b) $P_1/P_2 = 1000$; (c) $P_1/P_2 = 2000$; (d) $P_1/P_2 = \frac{1}{2}$; (e) $P_1/P_2 = \frac{3}{10}$.

---

† Decibel values are not quoted to more than two places of decimals since it is impracticable to measure to a greater accuracy. In many cases the practice of quoting decibels to the nearest tenth of a dB is sufficiently accurate.

Solution

(a) $P_1/P_2 = 2$, or in dB,

$$\frac{P_1}{P_2} = 10 \log_{10} 2 = 3 \text{ dB} \quad (Ans.)$$

(b) $P_1/P_2 = 1000$, or in dB,

$$\frac{P_1}{P_2} = 10 \log_{10} 1000 = 30 \text{ dB} \quad (Ans.)$$

(c) $P_1/P_2 = 2000$, or in dB,

$$\frac{P_1}{P_2} = 10 \log_{10} 2000 = 33 \text{ dB} \quad (Ans.)$$

(d) $P_1/P_2 = \frac{1}{2}$, or in dB,

$$\frac{P_1}{P_2} = 10 \log_{10} \tfrac{1}{2}$$
$$= 10 \log_{10} 1 - 10 \log_{10} 2 = 10 \times -0.3 = -3 \text{ dB} \quad (Ans.)$$

(e) $P_1/P_2 = \tfrac{3}{10}$, or in dB,

$$\frac{P_1}{P_2} = 10 \log_{10} \frac{3}{10}$$
$$= 10 \log_{10} 3 - 10 \log_{10} 10 = 10 \,(0.4771 - 1)$$
$$= -5.2 \text{ dB} \quad (Ans.)$$

Two things should be noted from Example 5.1. Firstly, a doubling, or halving, of power is equivalent to an increase, or decrease, of 3 dB. Thus, if a particular power ratio P_r/P_s is equivalent to 60 dB, then the ratio $2P_r/P_s$ is equivalent to 63 dB and the ratio $P_r/2P_s$ corresponds to 57 dB. Secondly, for power ratios of less than unity the method of calculation shown is straightforward and easy if the ratio is quoted as a fraction. Often, however, a power ratio of less than unity is quoted as a decimal when calculation will involve the use of bar numbers.

Now, it is true that $\log_{10} a/b = \log_{10} a - \log_{10} b$
$$= -(\log_{10} b - \log_{10} a)$$
$$= -\log_{10} b/a$$

and this relationship shows that the number of decibels corresponding to a particular power ratio can be calculated by *always* making the larger power the numerator in equation (5.1), and quoting the result as a gain if the output is the larger and as a loss if the input power is the larger. Bar numbers are avoided and the calculation is simplified.

EXAMPLE 5.2

Calculate the overall loss, or gain, in decibels of the arrangement shown in Fig. 5.2. If the input power is 10 mW calculate the output power.

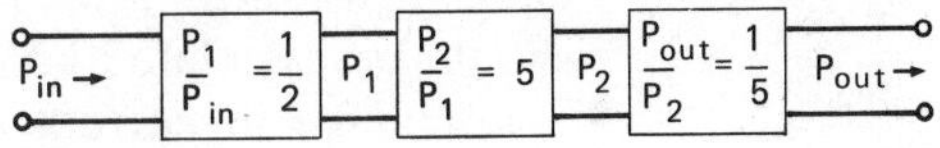

Fig. 5.2

$P_{in} \rightarrow \boxed{\dfrac{P_1}{P_{in}} = \dfrac{1}{2}} P_1 \boxed{\dfrac{P_2}{P_1} = 5} P_2 \boxed{\dfrac{P_{out}}{P_2} = \dfrac{1}{5}} P_{out} \rightarrow$

Solution
Loss of item 1 = 10 log$_{10}$ (P_{IN}/P_1) = 10 log$_{10}$ 2 = 3 dB
Gain of item 2 = 10 log$_{10}$ (P_2/P_1) = 10 log$_{10}$ 5 = 7 dB
Loss of item 3 = 10 log$_{10}$ (P_2/P_{OUT}) = 10 log$_{10}$ 5 = 7 dB

Overall loss = (7+3)−7, or 3 dB   (*Ans.*)
Therefore

$$10 \log_{10} \frac{P_{IN}}{P_{OUT}} = 3$$

$$\frac{P_{IN}}{P_{OUT}} = \text{antilog}_{10}\, 0.3 \simeq 2$$

and so the output power P_{OUT} is equal to $\tfrac{1}{2}P_{IN}$, or P_{OUT} = 5 mW   (*Ans.*)

Voltage and Current Ratios

A power ratio of x decibels is defined as

$$x = \log_{10} \frac{P_1}{P_2} \qquad (5.1)$$

The power P dissipated in a resistance R may be written $P = I^2 R$ or $P = V^2/R$, where I is the current flowing in the resistance and V is the voltage developed across the resistance. Hence equation (5.1) may be rewritten as either

$$x = 10 \log_{10} \frac{I_1^2 R_1}{I_2^2 R_2} \qquad \text{or} \qquad x = 10 \log_{10} \frac{V_1^2/R_1}{V_2^2/R_2}$$

$$= 10 \left[\log_{10} \frac{I_1^2}{I_2^2} + \log_{10} \frac{R_1}{R_2} \right] \qquad = 10 \left[\log_{10} \frac{V_1^2}{V_2^2} + \log_{10} \frac{R_2}{R_1} \right]$$

$$= 20 \log_{10} \frac{I_1}{I_2} + \log_{10} \frac{R_1}{R_2} \qquad = 20 \log_{10} \frac{V_1}{V_2} + \log_{10} \frac{R_2}{R_1}$$

If, *and only if*, $R_1 = R_2$ the resistances cancel and the equations become

$$x = 20 \log_{10} \frac{I_1}{I_2} \qquad (5.2)$$

or

$$x = 20 \log_{10} \frac{V_1}{V_2} \qquad (5.3)$$

Equations (5.2) and (5.3) may *only* be used when the resistances in which the currents I_1 and I_2 flow, or across which the voltages V_1 and V_2 are developed, are identical. When unequal resistances are involved the power dissipated in each should be calculated and equation (5.1) used.

A change in current or voltage at a point can always be quoted in dB, using either equation (5.2) or equation (5.3), since the same resistance is involved (provided, of course, that the resistance is not changed in value by the change in current or voltage).

EXAMPLE 5.3

An item of telecommunication equipment has an input resistance of 600 Ω and its output terminals are correctly terminated in a 600 Ω resistor. When a voltage of 1.5 V is applied across the input terminals a current of 15 mA flows in the load resistor. Calculate the loss, or gain, of the equipment.

Solution
Three methods of attack are possible: (*a*) calculate the input and output powers and use equation (5.1), (*b*) calculate the input current and use equation (5.2) and (*c*) calculate the output voltage and use equation (5.3). Methods (*b*) and (*c*) can be employed because the input and output powers are dissipated in equal resistances of 600 Ω.

Employing method (*c*),

Output voltage = $15 \times 10^{-3} \times 600 = 9$ V

Therefore

Gain of equipment = $20 \log_{10} 9/1.5 = 15.6$ dB   (*Ans.*)

EXAMPLE 5.4

An amplifier has a gain of 60 dB. If the input resistance of the amplifier is 75 Ω and its output terminals feed a matched load of 140 Ω, calculate the current flowing in the load when a voltage of 100 μV r.m.s. is applied to the input terminals.

Solution
The resistances in which the input and output powers are dissipated are unequal and hence equation (5.1) must be used.

$$\text{Input power to the amplifier} = \frac{(100 \times 10^{-6})^2}{75} = \frac{1 \times 10^{-8}}{75} \text{ watts}$$

Therefore

$$60 = 10 \log_{10} \frac{P_{OUT}}{(1 \times 10^{-8})/75}$$

$$\text{antilog}_{10} 6 = 75 P_{OUT} \times 10^8$$

$$1 \times 10^6 = 75 P_{OUT} \times 10^8$$

$$P_{OUT} = \frac{1}{7500} = I_{OUT}^2 \times 140$$

Therefore

$$I_{OUT} = \sqrt{\left(\frac{1}{7500 \times 140}\right)} = 0.976 \text{ mA} \quad (Ans.)$$

Reference Levels: The DBM, DBR and DBW

The decibel is not an absolute unit but is only a measure of a power *ratio*. It is meaningless to say, for example, that an amplifier has an output of 60 dB unless a reference level is quoted or is clearly understood. For example, a 60 dB increase on 1 microwatt gives a power level of 1 watt and a 60 dB increase on 1 watt gives a power level of 1 megawatt: here the same 60 dB difference expresses power differences of less than 1 watt in one case and nearly one million watts in the other. It is therefore customary in telecommunication engineering to express power levels as so many decibels above, or below, a clearly understood reference power level. This practice makes the decibel a more significant unit and allows it to be used for absolute measurements. The reference level most commonly used is 1 milliwatt and a larger power, P_1 watts, is said to have a level of

$$+x \text{ dBm}, \quad \text{where } x = 10 \log_{10} (P_1/1 \times 10^{-3})$$

and a smaller power, P_2 watts, is said to have a level of

$$-y \text{ dBm}, \quad \text{where } y = 10 \log_{10} (1 \times 10^{-3}/P_2).$$

EXAMPLE 5.5

Express in dBm the following power levels, (a) 1 watt, (b) 1 milliwatt and (c) 1 microwatt.

Solution

(a) $1 \text{ watt} = 10 \log_{10} \dfrac{1}{1 \times 10^{-3}} = 10 \times 3 = 30 \text{ dBm}$ (*Ans.*)

(b) $1 \text{ milliwatt} = 10 \log_{10} \dfrac{1 \times 10^{-3}}{1 \times 10^{-3}} = 10 \times 0 = 0 \text{ dBm}$ (*Ans.*)

(c) $1 \text{ microwatt} = -10 \log_{10} \dfrac{1 \times 10^{-3}}{1 \times 10^{-6}} = -10 \times 3 = -30 \text{ dBm}$ (*Ans.*)

In microwave radio-relay telephony/television systems a reference level of 1 watt is employed and power levels expressed in decibels relative to this level are quoted in dBW. A power level of 1 milliwatt is equal to $-10 \log_{10} (1/1 \times 10^{-3}) = -30$ dBW.

A further unit particularly useful in connection with multi-channel carrier telephony systems, is the dBr. This unit expresses in decibels the power level at a point, relative to the power level at some reference point. Normally the reference point is taken as the two-wire origin of a circuit.

80 THE DECIBEL

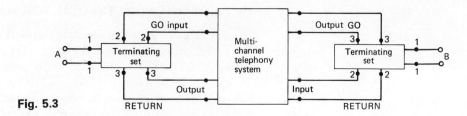

Fig. 5.3

EXAMPLE 5.6

Fig. 5.3 represents, in extremely simplified form, a four-wire telephony circuit that is routed over a multi-channel carrier-telephony system. It may be assumed that the loss through the terminating sets (these are devices for converting a two-wire line into a four-wire line) is 4 dB from terminals 1,1 to terminals 2,2; 4 dB from terminals 3,3 to terminals 1,1; and infinite from terminals 3,3 to terminals 2,2. The gain of the carrier system is 5 dB from input terminals to output terminals in both directions of transmission. Take, as is usual, the two-wire origin (point A) of the circuit as the reference point.

(a) If the power at point A is 0.25 mW what are the power levels in dBr at the input and output terminals of the carrier system and at the point B?
(b) What are the levels in dBm at the same points?
(c) What is the output power?

Solution
(a) Since the loss from terminals 1,1 to terminals 2,2 of a terminating set is 4 dB the power level at the GO input terminals of the carrier system is 4 dB below the power at the reference point A. Therefore

Level at GO input terminals = −4 dBr   (*Ans.*)

The gain of the carrier system is 5 dB. Therefore

Level at GO output terminals = +1 dBr   (*Ans.*)

The loss through a terminating set from terminals 3,3 to terminals 1,1 is 4 dB. Therefore

Level at point B = −3 dBr   (*Ans.*)

(b) The input power level in dBm is equal to $-10 \log_{10} \dfrac{1 \times 10^{-3}}{0.25 \times 10^{-3}}$ or −6 dBm. Therefore

The −4 dBr point has a level of −6 − 4 = −10 dBm   (*Ans.*)
The +1 dBr point has a level of −5 dBm   (*Ans.*)
The −3 dBr point has a level of −9 dBm   (*Ans.*)

(c) The output power level is −9 dBm. Therefore

$$9 = 10 \log_{10} \frac{1 \times 10^{-3}}{P_{OUT}}$$

$$\text{antilog}_{10} \, 0.9 = \frac{1 \times 10^{-3}}{P_{OUT}}$$

$$P_{OUT} = 0.125 \text{ mW} \quad (Ans.)$$

In part (c) the same answer can be obtained by noting that the point B is a −3 dBr point and remembering that −3 dB corresponds to a power ratio of one-half. Thus the output power is equal to one-half the input power.

The Decibel and The Human Ear

The human ear is capable of responding to a wide range of sound intensities and has a sensitivity which varies with change in frequency in a logarithmic manner. This makes the decibel a convenient unit for use with sound measurement and the measurement of sound equipment. Further, if the gain/frequency characteristic of an audio-frequency amplifier is to give a true indication of its aural effect it should consist of the gain expressed in decibels plotted to a base of frequency to a logarithmic scale.

The Neper

The neper is a logarithmic transmission unit which is widely used on the Continent and which expresses the ratio of two *currents*, or of two *voltages*, but *not* two powers as does the decibel.

A device is said to have a gain of x nepers if

$$x = \log_e \frac{I_{OUT}}{I_{IN}} \tag{5.4}$$

or

$$x = \log_e \frac{V_{OUT}}{V_{IN}} \tag{5.5}$$

where $I_{OUT} > I_{IN}$, $V_{OUT} > V_{IN}$, or a loss of y nepers if

$$y = \log_e \frac{I_{IN}}{I_{OUT}} \tag{5.4a}$$

$$y = \log_e \frac{V_{IN}}{V_{OUT}} \tag{5.5a}$$

where $I_{IN} > I_{OUT}$, $V_{IN} > V_{OUT}$.

It should be carefully noted that the neper is based on logarithms to the base e ($= 2.71828$) and *not* common (base 10) logarithms.

Relationship between the Decibel and the Neper

With a device or system having equal (matched) input and output impedances, a simple relationship exists between the loss or gain of the device or system quoted in nepers and quoted in decibels. Suppose a device has a loss of x nepers, then

$$x = \log_e \frac{I_{IN}}{I_{OUT}}$$

and

$$\frac{I_{IN}}{I_{OUT}} = e^x \quad \text{(from the definition of a logarithm)}$$

The loss y, in decibels, since impedances are equal, is

$$y = 20 \log_{10} \frac{I_{IN}}{I_{OUT}}$$

$$= 20 \log_{10} e^x = 20x \log_{10} e = 20x \times 0.4343 = 8.686x$$

Thus, in the particular case of matched impedances, 1 neper is equal to 8.686 dB.

Measurement of Decibels

A voltmeter can be calibrated to indicate decibel values directly. The voltmeter is used to measure the voltage developed across a resistance of known value, generally 600 Ω. The measured voltage V corresponds to a particular power and this can be expressed in decibels relative to 1 milliwatt. Thus

$$x \text{ dBm} = 10 \log_{10} \left[\frac{V^2/600}{1 \times 10^{-3}} \right] \tag{5.6}$$

Suppose a particular voltmeter has a voltage scale of 0–10 V and this scale is to be calibrated in dBm. Then

$$0 \text{ dBm} = 10 \log_{10} \left[\frac{V^2/600}{1 \times 10^{-3}} \right]$$

$$\frac{V^2}{600} = 1 \times 10^{-3}$$

i.e. $V = 0.775$ V.

A voltage of 1 volt corresponds to $10 \log_{10} \left[\frac{1/600}{1 \times 10^{-3}} \right]$ or +2.22 dBm,

2 volts gives $10 \log_{10} \left[\frac{4/600}{1 \times 10^{-3}} \right]$ or +8.24 dBm, and so on.

If the voltmeter is connected across a resistance of some other value than 600 Ω, the dBm value indicated by the meter will be incorrect. If, for example, the resistance was 1000 Ω and the voltage 2 V, then the true dBm value would be

$$10 \log_{10} \left[\frac{4/1000}{1 \times 10^{-3}} \right] \quad \text{or} \quad +6.02 \text{ dBm}$$

However, the value indicated by the voltmeter is +8.24 dB, i.e. 2.22 dB high. This error is equal to $10 \log_{10}[1000/600]$ dB. It can be concluded from this numerical example that the correction factor to be applied when the voltmeter is connected across a resistance R of other than 600 Ω is

$$-10 \log_{10}\left[\frac{R}{600}\right] \text{dB}$$

When the voltmeter is used to measure a voltage outside of the calibrated scale, another correction factor is required for correct dBm readings. Suppose the calibrated meter also has a 0–100 V scale with the scale markings for 10 V, 20 V, etc. coinciding with the markings 1 V, 2 V, etc. of the calibrated scale. If the pointer of the meter indicated 10 V it will lie on the +2.22 dBm mark. The true dBm reading is

$$10 \log_{10}\left[\frac{100/600}{1 \times 10^{-3}}\right] \quad \text{or} \quad 22.22 \text{ dB}$$

The indicated value is 20 dB low which is equal to $20 \log_{10}[100/10]$ dB. Thus, the correction factor required is

$$20 \log_{10}\left[\frac{\text{f.s.d. of scale used}}{\text{f.s.d. of calibrated scale}}\right] \text{dB}$$

EXAMPLE 5.7

A voltmeter has a 0–3 V scale which has been calibrated in dBm (0 dBm equal to 1 mW in 600 Ω). With the meter switched to its 0–60 V scale and connected across a 2000 Ω resistance the indicated dBm value is −3 dBm. Calculate the true dBm value.

Solution
The true dBm reading is

$$-3 - 10 \log_{10}\left[\frac{2000}{600}\right] + 20 \log_{10}\left[\frac{60}{3}\right] = -3 - 5.23 + 26.02$$
$$= +17.79 \text{ dBm} \quad (Ans.)$$

Exercises

5.1. Define the decibel and explain why it is a convenient unit for use in transmission problems.
   The input level to an amplifier is +24 dB relative to 1 μV and the amplifier has a gain of 30 dB. If the input and output impedances of the amplifier are equal and the output is matched to the load, calculate the input and output voltages.
   (C&G)

5.2. Define the decibel. Two amplifiers, having the gain/frequency responses given in the table below, are connected in tandem with a 10 dB resistive attenuator. Calculate and plot the overall gain of the combination, expressed in decibels, assuming that all input and output impedances are equal. (C&G)

Frequency (kHz)	60	66	72	78	84	90	96	102	108
Voltage gain, first amplifier	29.8	34.5	38.0	38.9	38.9	38.5	38.0	34.7	29.0
Voltage gain, second amplifier	28.2	37.2	37.6	36.7	36.7	36.7	38.7	39.1	29.8

5.3. Explain what is meant by a decibel.

The gain/frequency characteristics of two group amplifiers are given in the table.

Frequency (kHz)	60	76	92	108
Voltage gain of amplifier 1	310	330	340	390
Voltage gain of amplifier 2	330	345	380	325

If the two amplifiers are connected in tandem, separated by an attenuator having a loss of 15 dB, plot the overall gain/frequency characteristic of the combination, expressing the gain in decibels. It may be assumed that the amplifiers and the attenuator have the same input and output impedances.

(C&G)

5.4. Define the decibel and give three reasons why its use is convenient in transmission problems.

The input signal to an amplifier varies between 23.5 mW and 1.25 W. Express each power in dB relative to 1 mW and state the fluctuation in the level of the signal in dB. (C&G)

5.5. (a) State the advantages of using logarithmic units in transmission problems.

(b) Two amplifiers have the gain/frequency responses given in the table.

Frequency (kHz)	60	68	76	84	92
Voltage gain ratio in amplifier 1	305	325	335	340	345
Voltage gain ratio in amplifier 2	325	345	315	305	325

The amplifiers are connected in tandem separated by a 15 dB attenuator. Assuming that all input and output impedances are equal, calculate and plot the overall gain of the combination, expressed in decibels. (C&G)

5.6. A telecommunication circuit consists of four items of equipment connected in tandem by line and radio links as shown in Fig. 5.4.

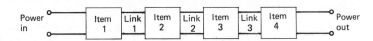

Fig. 5.4

The table below gives the input power to each item of equipment and the output power from items 1, 2 and 3. The power gain of item 4 is 23 dB.

	item 1	item 2	item 3	item 4
Power in (mW)	1000	316	500	251
Power out (mW)	25 100	12 600	15 800	

Determine: (a) the input power to item 1 in dBm, (b) the power gain in dB for each of items 1, 2 and 3, (c) the power loss in dB for each link, (d) the output power from item 4 in mW, (e) the overall power gain of the circuit in dB. (C&G)

5.7. Explain why logarithmic units are used for expressing the ratios of powers, currents and voltages in radio and line communication. Define the decibel. An amplifier has input and output load resistances of 600 Ω. The input signal voltage is +16 dB relative to 1 μV and the amplifier has a gain of 30 dB. Calculate: (a) the input and output voltages, (b) the output power. (C&G)

5.8. Define the decibel. Briefly explain why it is a convenient unit for use in conjunction with sound measurement and in transmission problems. Calculate the overall gain, or loss, in dB, of the arrangement shown in Fig. 5.5. If the input power is 30 mW, determine the output power in dBm. (C&G)

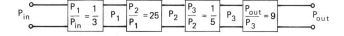

Fig. 5.5

5.9. (a) Give the meanings of the following terms: (i) = 20 dB with respect to 1 W, (ii) −10 dBW, (iii) +3 dBm, (iv) +12 dB with respect to 1 mV.

(b) The voltage gain of an amplifier is 26 times when terminated by an output resistance R Ω. This resistance is replaced by a 10 dB attenuator presenting the same input resistance R Ω and its output is terminated by another resistor R. Find: (i) the input voltage to the amplifier to give 50 mV across the output resistance of the attenuator, (ii) the power delivered to the amplifier when R is 600 Ω. (C&G)

86 THE DECIBEL

5.10. (*a*) Define the decibel and explain how it differs from other units used in telecommunications.
(*b*) A circuit of the same impedance throughout consists of two attenuators of 13 dB and 10 dB loss followed by an amplifier of 29 dB gain feeding a load resistor whose resistance is equal to the circuit impedance. An input of 1 V audio frequency is applied to the circuit. (i) Calculate the output voltage across the amplifier output resistance, (ii) calculate the a.c. power in the load resistor for a circuit impedance of 600 Ω.
(C&G)

5.11. (*a*) State the advantages of using logarithmic units in radio and line transmission work.

Fig. 5.6

(*b*) A telecommunication circuit consists of three items of equipment connected in tandem by line and radio links as shown in Fig. 5.6. The gains of items 1 and 3 are 23 dB and 16 dB respectively. The losses in links 1 and 2 are 30 dB and 42 dB respectively. If the input power to item 2 is 316 mW and the output from item 2 is 12600 mW determine, using decibel notation: (i) the gain of item 2, (ii) the output power from item 1, (iii) the input power to item 1, (iv) the input power to item 3, (v) the output power from item 3, (vi) the overall gain, or loss, of the circuit. (C&G)

5.12. (*a*) Explain how logarithmic units may be used in radio and line transmission work to simplify: (i) calculations, (ii) the presentation of data.

Fig. 5.7
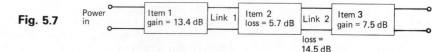

(*b*) A telecommunication circuit (Fig. 5.7) consists of three items of equipment connected in tandem by line and radio links. If the input power to item 1 is 161 mW and the output power from item 3 is 131 mW, determine in decibel notation: (i) the input powers to items 2 and 3, (ii) the output powers from items 1 and 2, (iii) the gain or loss in link 1. (C&G)

5.13. A wideband amplifier having identical input and output impedances gives output voltages at various frequencies as shown in the table. This table also shows the increased output in dB obtained by switching into the amplifier a high-frequency boost circuit.

Frequency (kHz)	10	100	500	1000	2000	3000	5000	8000
Output voltage (V)	3.5	3.75	3.9	3.7	3.1	2.4	1.4	0.4
High-frequency boost (dB)	0	0	0.5	1.5	3.9	6.2	8.1	9.9

(a) Plot against a logarithmic frequency scale, the curve of voltage gain in dB relative to a constant input voltage of 0.5 V for the amplifier (i) without boost, (ii) with boost applied.

(b) Estimate from these graphs (i) the output voltage at 1.5 MHz with boost switched in, (ii) the frequency at which, without boost, the amplifier provides unity gain, (iii) the extended frequency, due to boost, before the voltage output is 3 dB down on its value at 100 kHz. (C&G)

5.14. (a) State the advantages of expressing power ratios in decibels.

(b) Why may it be advantageous for the gain control of an a.f. amplifier to obey a logarithmic law?

(c) The gain control of an amplifier is graduated with numbers 1 to 5 at equally spaced intervals. With a constant input signal the output power of the amplifier varies with gain setting in accordance with the table. (i) Plot a graph of output power

Gain control step	1	2	3	4	5
Output power (mW)	6.31	39.8	251	1590	10 000

against maximum output power expressed in dB, using the gain control steps as a base. (ii) What is the change in output, expressed in dB, between step 2 and step 3? (C&G)

5.15. A voltmeter has a 0–10 V scale calibrated to read dBm relative to 600 Ω. When the meter is connected across a 100 Ω resistor and switched to the 0–1 V scale, the indicated dBm value is +4.5 dB. Calculate the true dBm value.

Short Exercises

5.16. Express the following power ratios in dB; 4, 8, 16, 100, 200 and 10^4.

5.17. Express the following current ratios in dB; 4, 8, 16, 100, 200 and 10^4. State any assumption made.

5.18. The input power to an amplifier is 50 W. Calculate the output power if the gain of the amplifier is (i) 10 dB, (ii) 20 dB, (iii) 23 dB, (iv) 26 dB and (v) 40 dB.

5.19. An amplifier has a gain of 33 dB. Calculate the power delivered to its input terminals if the power output of the amplifier is (i) 25 mW, (ii) 50 mW and (iii) 2 W.

5.20. The output power of an amplifier is (i) 50 mW, (ii) 200 mW, (iii) 1 W and (iv) 5 W. Express these power levels in dBm.

5.21. The input power to an amplifier is (i) −10 dBm, (ii) −4 dBm, (iii) 0 dBm and (iv) +10 dBm. If the gain of the amplifier is 20 dB calculate the output power in (a) dBm and (b) watts.

5.22. An amplifier has a gain of 56 dB. Calculate its input power when the power output is (i) 10 W, (ii) 2 W, (iii) 0.5 W.

5.23. An attenuator has a loss of 9 dB. A power of 250 mW is applied across its input terminals. Calculate the output power.

5.24. The input power to 5 km length of telephone cable is 100 mW. If the output power is 8 mW what is the loss of the cable per km?

5.25. Express the following power ratios in dB: 2, 4, 10, 100, 1000.

6 Telephone Lines and Cables

The telephone network of Great Britain is divided into local lines, junctions and trunks. Local lines are unamplified circuits that connect the individual telephone subscribers to their local telephone exchanges; junctions are two-wire circuits that may or may not be amplified and connect nearby telephone exchanges together; and trunks are four-wire circuits that interconnect more distant telephone exchanges. Most local lines and junctions employ pairs in audio-frequency telephone cables, although pulse-code-modulated multi-channel systems are now being introduced into the junction network. Open-wire lines are frequently used for the final distribution of the local lines to the subscribers' premises and, rarely nowadays, for some junction and trunk circuits.

Most trunk circuits are routed wholly or partly over multi-channel frequency-division telephony systems (described in Chapter 8). These systems, in turn, are routed over either star-quad carrier cables or coaxial telephony systems using either coaxial tubes or microwave radio-relay systems. Both two-wire lines and coaxial tubes are also used as feeders in radio stations to connect transmitters and/or receivers to aerials.

Basic Transmission Line Theory

A transmission line consists of a pair of conductors separated from each other by a dielectric. Two main types of line exist: the two-wire line, shown in Fig. 6.1a and the coaxial line shown at 6.1b. The two-wire line may be either open-wire or a cable pair. The coaxial pair is nearly always operated with the outer conductor earthed since the outer then acts as an efficient screen at all operating frequencies, and is said to be *unbalanced*.

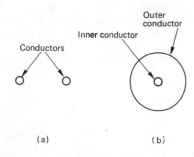

Fig. 6.1 (a) The two-wire pair (b) the coaxial pair

TELEPHONE LINES AND CABLES

The conductors forming a pair have both *resistance* and *inductance* uniformly distributed along their length and uniformly distributed *capacitance* and *leakance* between them. These four quantities are known as the *primary coefficients* of a line.

The Primary Coefficients of a Line

RESISTANCE

The resistance R of a unit length of line, or *loop* resistance, is the sum of the resistances of the two conductors comprising a pair. The unit length of a line is the metre.

At zero frequency the resistance of a line is the d.c. resistance R_{dc} given by

$$R_{dc} = \frac{\rho_1}{a_1} + \frac{\rho_2}{a_2} \quad \text{ohms per loop metre} \tag{6.1}$$

where ρ_1 and ρ_2 are the resistivities of the two conductors, and a_1 and a_2 are their cross-sectional areas.

At a frequency of few kilohertz or so, a phenomenon known as *skin effect* comes into play and causes current to flow only in a thin layer or "skin" at the outer surface of the conductor. The higher the frequency the thinner this skin becomes and the smaller the cross-section of the conductor in which the current flows. Since the resistance is inversely proportional to the cross-sectional area of the "effective" conductor, the *a.c. resistance* increases with increase in frequency. When skin effect is fully developed the a.c. resistance is proportional to the square root of the frequency, i.e.

$$R_{ac} = k_1 \sqrt{f} \tag{6.2}$$

where k_1 is a constant.

Fig. 6.2 shows how R_{ac} varies with frequency. Initially little variation from the d.c. value is observed, but at higher frequencies the relationship given in equation (6.2) is true.

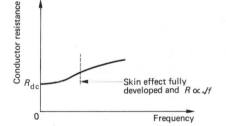

Fig. 6.2 Resistance/frequency characteristic of a transmission line

INDUCTANCE AND CAPACITANCE

The loop inductance L and the shunt capacitance C of a line, in henrys per loop metre and farads per metre respectively, are both more or less constant with change in frequency.

LEAKANCE

The leakance G of a line in siemens per metre represents the leakage of current between the conductors via the dielectric separating them, and is the reciprocal of insulation resistance. The leakage current has two components: one passes through the insulation between the conductors, and the other supplies the power losses in the dielectric itself as the line capacitance

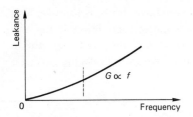

Fig. 6.3 Leakance/frequency characteristic of a transmission line

is charged and discharged. Leakance increases with increase in frequency and at the higher frequencies it is directly proportional to frequency, i.e.

$$G = k_2 f \tag{6.3}$$

where k_2 is another constant. Fig. 6.3 shows how the leakance of a pair of conductors varies with change in frequency.

The Secondary Coefficients of a Line

The secondary coefficients of a transmission line are its *characteristic impedance*, its *attenuation coefficient*, its *phase-change coefficient* and its *velocity of propagation*.

CHARACTERISTIC IMPEDANCE

The characteristic impedance Z_0 of a transmission line is the input impedance of an infinite length of that line. Fig. 6.4 shows an infinite length of line; its input impedance is the ratio of the voltage V_s impressed across the sending-end terminals to the current I_s flowing into the line, i.e.

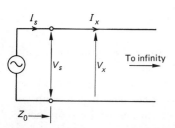

Fig. 6.4 Definition of the characteristic impedance of a line

$$Z_0 = \frac{V_s}{I_s} \text{ ohms} \tag{6.4}$$

Similarly, at any point x along the line, the ratio V_x/I_x is always equal to Z_0.

Suppose the line is now cut a finite distance from its sending-end terminals as shown in Fig. 6.5a. The remainder of the line is still of infinite length and so the impedance measured at terminals 2–2 is equal to the characteristic impedance. Thus before the line was cut, terminals 1–1 were effectively terminated in impedance Z_0. The conditions at the input terminals will not be changed if terminals 1–1 are closed in a physical impedance equal to Z_0, as in Fig. 6.5b. This leads to a more practical definition: the characteristic impedance of a transmission line is the input impedance of a line that is itself terminated in the characteristic impedance.

A line that is terminated in its characteristic impedance is said to be *correctly terminated*.

The characteristic impedance of a line depends upon the values of the primary coefficients of that line, according to the equation

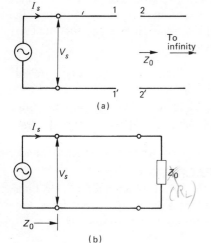

Fig. 6.5 Alternative definition of the characteristic impedance of a line

$$Z_0 = \sqrt{\frac{R + j\omega L}{G + j\omega C}} \text{ ohms} \tag{6.5}$$

At higher frequencies where $\omega L \gg R$ and $\omega C \gg G$ equation (6.5) reduces to the more convenient form

$$Z_0 = \sqrt{\frac{L}{C}} \text{ ohms} \tag{6.6}$$

This equation always applies to coaxial lines since they are only operated at frequencies high enough to make R and G negligible with respect to ωL and ωC respectively.

EXAMPLE 6.1

A generator of e.m.f. 1 V and internal impedance 79 Ω is applied to a line having $L = 0.5$ mH/km and $C = 0.08$ μF/km. If the approximate expression, $Z_0 = \sqrt{(L/C)}$ ohms, for characteristic impedance may be assumed, calculate (a) the sending-end current, and (b) the sending-end voltage.

Solution

$$Z_0 = \sqrt{\frac{0.5 \times 10^{-3}}{0.08 \times 10^{-6}}} = 79 \ \Omega$$

Hence, referring to Fig. 6.6,

$$I_s = \frac{1}{79+79} = \frac{1}{158} \text{ A} \approx 6.33 \text{ mA} \qquad (Ans. \ (a))$$

$$V_s = 79 I_s = \frac{79 \times 1}{158} = 0.5 \text{ V} \qquad (Ans. \ (b))$$

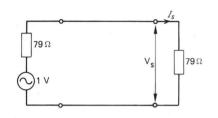

Fig. 6.6

Analysis beyond the scope of this book shows that the characteristic impedance of a transmission line is a function of the dimensions of the line and the permittivity of the dielectric.

For an air-spaced coaxial line the characteristic impedance is given by

$$Z_0 = 138 \log_{10} \frac{R}{r} \text{ ohms} \qquad (6.7)$$

where R is the inner radius of the outer conductor, and r is the radius of the inner conductor.

The characteristic impedance of a coaxial line is confined somewhere in the range 30–100 Ω, because of construction difficulties arising from the need for either an excessively thin inner conductor or a very small spacing between inner and outer. Most practical cables have a characteristic impedance of 50–75 Ω.

For an air-spaced two-wire line,

$$Z_0 = 276 \log_{10} \frac{D}{r} \text{ ohms} \qquad (6.8)$$

where D is the spacing between the centres of the two conductors and r is the radius of each conductor. Z_0 is generally some hundreds of ohms.

If there is a continuous insulation between the conductors the characteristic impedance is reduced from Z_0 to $Z_0/\sqrt{\epsilon_r}$, where ϵ_r is the relative permittivity of the dielectric.

EXAMPLE 6.2

A flexible coaxial feeder has an outer conductor of diameter 8.48 mm and an inner conductor of diameter 1.42 mm. If the cable has a continuous polythene dielectric of relative permittivity 2.3, calculate its characteristic impedance.

Solution
From equation (6.7),

$$Z_0 = \frac{138}{\sqrt{2.3}} \log_{10} \frac{8.48}{1.42}$$

$$= 70.62 \, \Omega \quad (Ans.)$$

ATTENUATION COEFFICIENT

As a current or voltage is propagated along a line its amplitude is progressively reduced or *attenuated* because of losses in the line. These losses are of two types: firstly, conductor losses caused by I^2R power dissipation in the series resistance, and secondly, dielectric losses. If the current or voltage at the sending-end terminals of the line is I_s, or V_s, then the current, or voltage, at one metre distance along the line is $I_1 = I_s e^{-\alpha}$, or $V_1 = V_s e^{-\alpha}$, where e is the base of the natural logarithms (2.7183) and α is the *attenuation coefficient* of the line. In the next metre distance the attenuation is the same, and thus the current I_2 at the end of this distance is

$$I_2 = I_1 e^{-\alpha} = I_s e^{-\alpha} e^{-\alpha} = I_s e^{-2\alpha}$$

If the line is l metres long, the received current and voltage are given, respectively, by

$$I_r = I_s e^{-\alpha l} \quad (6.9)$$

$$V_r = V_s e^{-\alpha l} \quad (6.10)$$

Thus both current and voltage waves decay exponentially as they are propagated along the line (Fig. 6.7).

The general expression for the attenuation coefficient of a line is complex and beyond the scope of this book. However, at frequencies where $\omega L \gg R$ and $\omega C \gg G$ the expression simplifies to

$$\alpha = \frac{R}{2Z_0} + \frac{GZ_0}{2} \quad \text{nepers per metre} \quad (6.11)$$

The characteristic impedance Z_0 of a line is given by $\sqrt{(L/C)}$ when $\omega L \gg R$ and $\omega C \gg G$ and is therefore constant with change in frequency, and consequently the attenuation coefficient α varies with frequency in accordance with the frequency dependencies of R and G, i.e.

$$\alpha = k_3 \sqrt{f} + k_4 f \quad \text{nepers per metre} \quad (6.12)$$

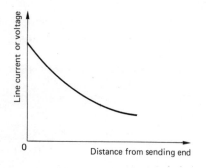

Fig. 6.7 Decay of current and voltage along a transmission line

$I_x = I_s e^{-\alpha x}$ or $V_x = V_s e^{-\alpha x}$

Generally, in Great Britain and the United States, the attenuation coefficient is expressed in decibels instead of nepers, the relationship being

$$1 \text{ Np} = 8.686 \text{ dB} \quad \text{(See Chapter 5)}$$

EXAMPLE 6.3

A coaxial cable has a loss of 3.5 dB/km at 1 MHz. Calculate its loss at 4 MHz if (a) the dielectric loss is negligible, and (b) the dielectric loss is 10% of the total.

Solution

(a) Loss at 4 MHz = loss at 1 MHz $\times \sqrt{\dfrac{4 \times 10^6}{1 \times 10^6}}$

$= 3.5 \times 2 = 7 \text{ dB/km} \quad (Ans. (a))$

(b) Dielectric loss at 1 MHz = 0.35 dB/km
Conductor loss at 1 MHz = 3.15 dB/km

Loss at 4 MHz = $3.15 \times \sqrt{\dfrac{4}{1}} + 0.35 \times \dfrac{4}{1}$

$= 6.3 + 1.4 = 7.7 \text{ dB/km} \quad (Ans. (b))$

PHASE-CHANGE COEFFICIENT

A current or voltage wave travels along a line with a finite velocity and so the current, or voltage, at the end of a metre length of line lags the current, or voltage, entering that length. The phase difference between the line currents, or voltages, at two points which are one metre apart is known as the *phase-change coefficient* β of the line. β is measured in radians per metre. In each metre distance of a line the same phase shift is introduced; consequently for a line l metres in length the received current will lag the sending-end current by βl radians.

EXAMPLE 6.4

A correctly terminated transmission line has $Z_0 = 500\ \Omega$, $\alpha = 1$ dB/km and $\beta = 30°$/km, and is 3 km long. A $500\ \Omega$ source, of e.m.f. 2 V, is applied to the sending-end terminals of the line. Calculate (a) the magnitude of the received current, and (b) its phase relative to the sending-end voltage.

Solution
Since the line is correctly terminated its input impedance is equal to its characteristic impedance of $500\ \Omega$. Therefore

$$I_s = \frac{2}{500 + 500} = 2 \text{ mA}$$

Line loss = $3 \times 1 = 3$ dB

The load and input impedances of the line are both $500\ \Omega$ and so use may be made of the expression

Attenuation in decibels = $20 \log_{10}$ (current ratio)

Thus

$$3 = 20 \log_{10} \frac{2}{I_r}$$

$$\frac{2}{I_r} = \text{antilog}\frac{3}{20} = 0.1413 = \sqrt{2}$$

and

$$I_r = \frac{2}{\sqrt{2}} = \sqrt{2} \text{ mA} \quad (Ans.\ (a))$$

The phase shift introduced by the line is $3 \times 30° = 90°$; therefore

I_r lags V_s by $90°$ $\quad (Ans.\ (b))$

PHASE VELOCITY OF PROPAGATION

The phase velocity v_p of a line is the velocity with which a sinusoidal wave travels along that line. Any sinusoidal wave travels with a velocity of one wavelength per cycle. There are f cycles per second and so a wave travels with a velocity of λf metres per second, i.e.

$$v_p = \lambda f \quad \text{metres per second} \tag{6.13}$$

where λ is the wavelength and f is the frequency of the sinusoidal wave.

In one wavelength a phase change of 2π radians occurs, and hence the phase change per metre is $2\pi/\lambda$ radians, and this is also equal to the phase-change coefficient. Thus

$$\beta = \frac{2\pi}{\lambda} \quad \text{or} \quad \lambda = \frac{2\pi}{\beta} \tag{6.14}$$

and

$$v_p = \frac{2\pi}{\beta} \times f = \frac{\omega}{\beta} \quad \text{metres per second} \tag{6.15}$$

Group Velocity

Any repetitive, non-sinusoidal waveform contains components at a number of different frequencies, each of which will be propagated along a transmission line with a phase velocity given by equation (6.15). For all these components to travel with the same velocity and arrive at the far end of the line at the same moment, it is necessary for the phase change coefficient β of the line to be a linear function of frequency, i.e. for ω/β to be a constant at all frequencies. It is only at radio frequencies that practical lines satisfy this requirement. At lower frequencies β varies with frequency in a non-linear manner. Fig. 6.8 shows the β-frequency characteristic of a typical audio-frequency cable.

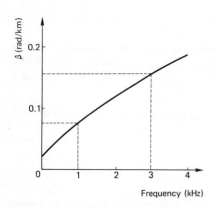

Fig. 6.8 Phase-change coefficient/frequency characteristic of a transmission line

Suppose a signal consisting of a 1000 Hz fundamental plus 50% third harmonic is applied to a 10 km length of a pair in this cable. The fundamental frequency component will be propagated with a phase velocity

$$\frac{\omega}{\beta} \text{ of } \frac{2\pi \times 1000}{0.075 \times 10^{-3}} \text{ or } 83.76 \times 10^6 \text{ m/s}$$

and the component at the third harmonic will propagate with a phase velocity of

$$\frac{\omega}{\beta} = \frac{6\pi \times 1000}{0.155 \times 10^{-3}} \text{ or } 121.61 \times 10^6 \text{ m/s}$$

This means that the harmonic component will arrive at the far-end of the line t seconds before the fundamental arrives.

$$\text{Time } t = \frac{\text{Length of line}}{\text{Velocity difference}} = \frac{10^4}{(121.61 - 83.76) \times 10^6}$$
$$= 0.264 \text{ ms}$$

This time is 0.264 times the periodic time of the fundamental frequency component and so the third harmonic component has a phase lead of $0.264 \times 360°$ or approximately 95°. Figs. 6.9a and b show, respectively, the resultant waveforms at the beginning and at the end of the line; it is evident that waveform distortion has taken place.

It is customary to consider the *group velocity* of a complex wave rather than the phase velocities of its individual frequency components. Group velocity is the velocity with which the *envelope* of the resultant waveform is propagated. Fig. 6.10 for example illustrates the meaning of the term group velocity when it is applied to the transmission of an amplitude-modulated wave over a line. The envelope travels at the group velocity, while the carrier, which is one of the component frequencies of the modulated wave, propagates with its particular phase velocity.

If a narrow band $\omega_2 - \omega_1$ of frequencies is transmitted over a line and at these two frequencies the phase change coefficients of the line are β_2 and β_1 respectively, then the group velocity V_g is given by equation (6.16), i.e.

$$V_g = \frac{\omega_2 - \omega_1}{\beta_2 - \beta_1} \text{ m/s} \tag{6.16}$$

The *group delay* of a line is the product of the length of the line and reciprocal of its group velocity.

EXAMPLE 6.5

A signal containing components at 1 kHz and at 2 kHz is propagated along a line. At 1 kHz the phase change coefficient β is 0.07 rad/km

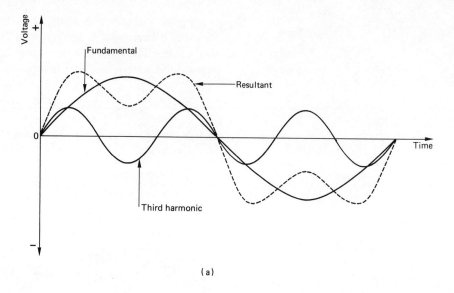

(a)

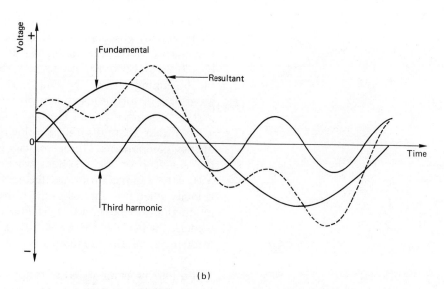

(b)

Fig. 6.9 Showing the envelope of a fundamental and its third harmonic (a) at the beginning and (b) at the end of a line in which the ratio ω/β is not constant

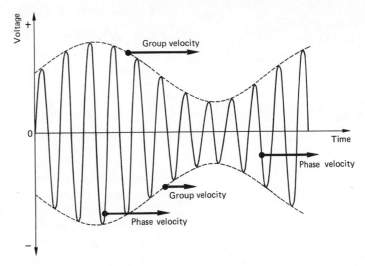

Fig. 6.10 Group and phase velocities of an amplitude-modulated wave

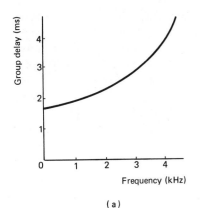

(a)

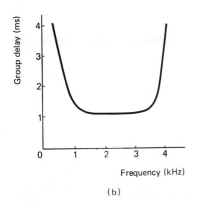

(b)

Fig. 6.11 Group-delay/frequency characteristics of (a) an audio-frequency cable and (b) a channel in a multi-channel telephony group

and at 2 kHz, β is 0.14 rad/km. Calculate the group velocity of the signal. Determine also its group delay if the line is 10 km in length.

Solution
From equation (6.16)

$$V_g = \frac{2\pi(2000-1000)}{(0.14-0.07)\times 10^{-3}} = 89.76\times 10^3 \text{ km/s} \quad (Ans.)$$

The group delay is

$$\frac{1}{89.76\times 10^3}\times 10 = 0.11 \text{ ms} \quad (Ans.)$$

Two typical group delay/frequency characteristics are given in Fig. 6.11. Fig. 6.11a shows clearly that the group delay on an audio-frequency cable increases with increase in frequency; obviously it will also increase as the length of the line is increased. The group delay/frequency characteristic of a channel in a multi-channel telephony system increases at both low and high frequencies because of the presence of the channel filters. When the group delay of a circuit varies with frequency, the different components of a complex signal will not arrive at the end of the line at the same time and *group delay distortion* will be present.

Construction of Telephone Cables

The material used for the conductors in a telephone cable must satisfy several requirements. It must have high electrical conductivity, it must be as light as possible, it must have adequate tensile strength, it must be ductile, and it must be cheap. For many years copper has been the material used, but in recent years the price of copper has increased to such an extent that consideration has been given to aluminium as the conductor material. The conductivity of aluminium is approximately 60% of that of copper and the specific gravity of aluminium is about 30% of that of copper. For a given resistance, therefore, an aluminium conductor has four-thirds the diameter of a copper conductor and is only half as heavy. The increased diameter means that greater quantities of insulating and sheath material are required, and that a given duct space will accommodate fewer cables. Even so, economies are still to be expected.

The traditional material for cable sheaths has been lead because it has the advantages of flexibility, ease of application and jointing, and mechanical strength. Nowadays many cable sheaths are made of polythene. Polythene is a good insulator, it is resistant to abrasion, water and corrosion. Polythene-sheathed cables are cleaner and lighter than lead and this permits longer lengths of cable to be inserted into ducts, reducing the number of joints required. A considerable reduction in installation costs is thus possible. The disadvantages of a polythene sheath are that it allows the ingress of water vapour and is easily damaged by applied heat.

External Cables

(1) Audio-Frequency Cables

The cables used for junction and trunk circuits are of the paper-core star-quad type. Star quad means that the cable is made up by grouping the conductors in fours, or quads. The arrangement of a single quad is shown in Fig. 6.12.

The conductors are made of annealed copper and are insulated with a helical lapping of paper string over which there is a lapping of paper; this arrangement ensures that there is a layer of air between conductor and paper. Air is a good insulator of low permittivity, and relying on air to provide a large proportion of the cable insulation ensures low capacitance between conductors. The paper lappings are printed with identifying inked rings; to maintain the same capacitance between the four conductors of a quad it is necessary to ensure that the quantity of ink on the insulation of all four conductors

Fig. 6.12 Construction of an audio-frequency star quad

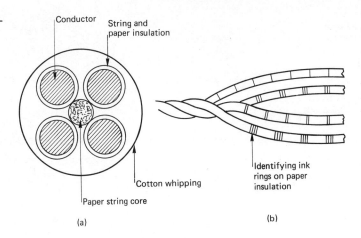

(a) (b)

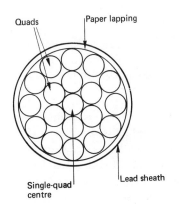

Fig. 6.13 Construction of an audio-frequency star quad cable

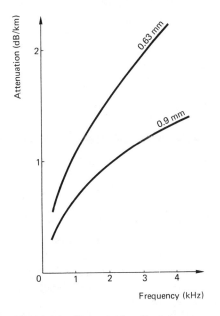

Fig. 6.14 Attenuation/frequency characteristics of audio-frequency star quad cable

is the same. This is achieved by staggering the spacings of the inked rings as shown in Fig. 6.12b. The four insulated conductors are then twisted around a paper string core to form a quad. Opposite conductors comprise a pair. A whipping of coloured cotton thread is then placed around each quad, the colour identifying the quad, and a number of quads are stranded in layers to form the cable core.

There are three methods of combining the quads, employing, respectively, one, three or four quads as the centre core. Fig. 6.13 shows the arrangement using a single-quad central core. To reduce interference between quads the direction of stranding† is changed in successive layers and four different lays of twist lengths‡ are used. The complete core is lapped with two or more paper layers and then the lead sheath is applied.

Audio-frequency star-quad cables are nowadays made in sizes ranging from 14 to 1040 pairs using 0.63 mm copper conductors; and also 14 to 504 pairs using 0.9 mm copper conductor size are used. The characteristic impedance of this type of cable is 600–800 Ω at low audio frequencies and 200–300 Ω at 3000 Hz; the attenuation is shown in Fig. 6.14 for both 0.63 mm and 0.9 mm conductors.

For local line cables of more than 100 pairs the polythene-sheathed paper-core unit twin cable is now standard in the BPO network. In a unit twin cable the conductors are twisted together in pairs and then 50 pairs at a time (or 25 pairs or more often 100 pairs) are grouped together to form a unit. A number of units, according to the size of the cable, are then combined to form the cable as shown in Fig. 6.15. The

† Axial length in which a quad makes one complete twist.
‡ Axial length in which a layer makes one complete rotation.

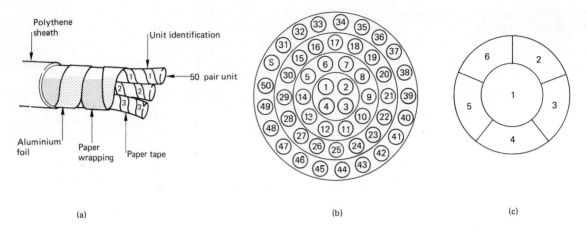

Fig. 6.15 Construction of an audio-frequency unit-twin cable

conductors are lapped in paper tape which is printed with identifying coloured ink rings. The conductors are then twisted together in pairs with different twist lengths for adjacent pairs. A number of pairs are then grouped together to form a self-contained unit and each unit is wrapped in paper tape and whipped with cotton. Layers within a unit are separated by cotton whippings and each complete unit is wrapped with two layers of insulating paper. Over this is wound an aluminium tape, the outside of which has been coated with a thin film of polythene. The polythene sheath is then applied on top of the polythene film. The aluminium tape prevents moisture penetrating the sheath and reaching the paper core. There is a variety of different cable sizes in current use, such as, for example, 400 pairs with 0.32 mm copper conductors and 800 pairs with 0.63 mm conductors. Fig. 6.15b shows the cross-section of a 50 pair unit and Fig. 6.15c how six such units are combined to produce a 300 pair cable.

The standard local cable for up to 100 pairs employs polythene as both the sheath and the conductor insulation material. Cables with only two pairs use the star quad arrangement, but all the larger sizes have conductors twisted in pairs. The pairs are then stranded in layers around a central core that comprises one, two, three or four pairs. For example, a 50-pair cable consists of three centre pairs, nine first-layer pairs, 16 second-layer pairs and 22 outer-layer pairs. The core is then lapped with two layers of paper before the polythene sheath is applied. The polythene contains carbon black in order to minimize deterioration by sunlight.

All polythene cables can be directly terminated onto terminal blocks but require a water barrier at each joint to prevent water entering the cable passing into adjacent jointing lengths.

Cable conductor diameters are 0.4 mm, 0.5 mm, 0.63 mm and 0.9 mm. Some modern cables are provided with 0.5 mm or 0.8 mm aluminium conductors.

(2) Coaxial Cables

The cables previously described in this chapter cannot be used for the transmission of signals lying much above the audio-frequency range because of excessive attenuation. The high attenuation is due mainly to the relatively high capacitance between conductors. Star-quad cables exist in which low capacitance is obtained by wrapping the insulating paper more loosely around the conductors to trap more air between conductors; these cables can be used up to about 250 kHz. Twenty-four-channel telephony systems operate using this type of cable in the frequency band 12–108 kHz, the characteristic impedance of the cable being 140 Ω.

Most multi-channel line telephony systems are wideband systems that employ coaxial cables as the transmission medium. Two main types of coaxial cable are in use in Great Britain, namely the 2.6/9.5 and 1.2/4.4 cables (the types are known by the diameter of the inner and outer conductors in millimetres).

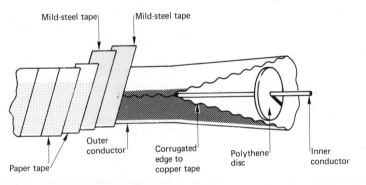

Fig. 6.16 Construction of a 2.6/9.5 coaxial tube

The construction of a 2.6/9.5 coaxial pair is shown in Fig. 6.16. The inner conductor is of solid copper 2.6 mm in diameter, and has polythene spacing discs mounted on it at 33 mm intervals. The concentric outer conductor is made from copper tape folded into tubular shape, and having corrugated edges to ensure that one side of the tape does not ride over the other. The inner diameter of the outer conductor is 9.5 mm and this results in the cable having a characteristic impedance of 75 Ω. Two mild-steel tapes, 0.127 mm thick, are wound over the outer conductor to give it mechanical strength and also to screen the pair against low-frequency interference. At high

102 TELEPHONE LINES AND CABLES

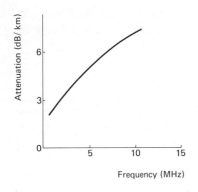

Fig. 6.17 Attenuation/frequency characteristics of a 2.6/9.5 coaxial tube

frequencies a coaxial pair is self-screening because the inner conductor is completely enclosed. The attenuation/frequency characteristic of this type of coaxial pair is shown in Fig. 6.17.

A coaxial cable consists of two, four, six or eight coaxial pairs together with a number of star quads fitted into the interstices. Fig. 6.18 shows the make-up of a four-tube cable; the groups of interstice quads are each wrapped in two layers of paper. The complete cable core is wrapped with two layers of paper tape and then the lead sheath is applied.

A new type of coaxial cable contains eighteen 2.6/9.5 mm coaxial pairs with a polythene-covered lead sheath. The coaxial pairs are positioned within the sheath in two layers, the inner layer containing 6 pairs and the outer layer 12 pairs. Each of the outer interstices has a 0.63 mm quad cable situated within it. An end view of the cable construction is shown in Fig. 6.19.

A 1.2/4.4 coaxial pair has an outer conductor of 4.4 mm inner diameter, and an inner conductor of 1.2 mm diameter, these dimensions giving the pair a characteristic impedance of 75 Ω. There are several different methods of construction, and one of them is shown in Fig. 6.20. The construction is similar to that of a 2.6/9.5 coaxial pair, but a lapping of thin polythene tape is provided between the polythene spacing discs and the outer conductor. One, two, three, four, six or eight coaxial pairs together with a number of quads in the interstices form a complete cable.

2.6/9.5 type coaxial cables are used for long-distance coaxial telephony systems that satisfy CCITT recommendations, and are capable of handling up to 10 800 telephony channels in the frequency band 4404–59 580 kHz.

1.2/4.4 type coaxial cables—known as "small-diameter coaxial cables"—are used for short-distance coaxial telephony systems.

Flexible coaxial cables are often employed in v.h.f. and u.h.f. radio and television systems; for example, as a means of connecting the TV receiving aerial on the rooftop, or in the loft, to the TV receiver in the home. The inner conductor is either a solid copper wire or a number of copper strands, and it is surrounded by a flexible insulating material such as polythene. The outer conductor consists of a copper braid wound around the insulation, and over this is a flexible covering such as vinyl plastic.

The outer covering excludes moisture and air and thus prevents oxidation and corrosion of the copper braid, and also protects the braid from mechanical damage. Fig. 6.21 shows the construction of a flexible coaxial cable. Typically such a cable may have the dimensions given in Example 6.2, leading to a characteristic impedance of about 70–75 Ω.

TELEPHONE LINES AND CABLES 103

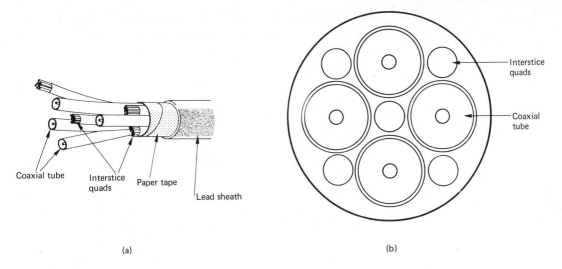

Fig. 6.18 Construction of 4-tube cable

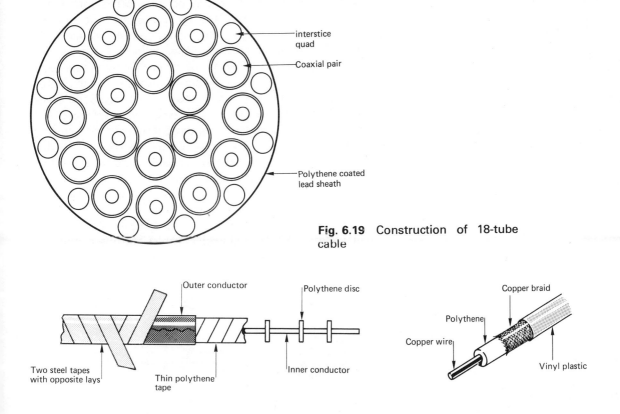

Fig. 6.19 Construction of 18-tube cable

Fig. 6.20 Construction of 1.2/4.4 coaxial tube

Fig. 6.21 Construction of a flexible coaxial cable

The attenuation of a flexible coaxial cable is also a function of frequency; for example, the cable of Example 6.2 has a loss of approximately 0.2 dB/m at 30 MHz rising to about 0.8 dB/m at 300 MHz.

Submarine Cables

Cables laid in the shallow seas near land must be adequately protected against damage caused by the abrasive action of tides, rocks, anchors, fishing trawls, etc. The necessary protection is given by steel armour wires around the cable. Armoured cable suffers from two disadvantages: (*a*) it is costly, and (*b*) it is extremely difficult to avoid twists and kinks appearing in the cable whilst laying it on the sea bed or during its recovery for repair. Such twists and kinks can damage the cable and are therefore undesirable. In deep water a cable is undisturbed and there is no need to provide armoured protection, and for the CANTAT system a new "lightweight" submarine cable was introduced (Fig. 6.22).

A centre steel strand consisting of a number of high-tensile steel wires is placed inside the inner conductor to give the cable the required strength. The inner conductor is made of a longitudinal copper tape bent into cylindrical shape, and around it is extruded a layer of polythene. Six aluminium tapes are applied with a lay of 40.64 cm to form the outer conductor, and a polythene tape is wound over the tapes. Some layers of aluminium tape are then interleaved and overlapped with polythene tape to form a screen, these layers are then covered by cotton tape, and finally the polythene sheath is applied.

The type of armoured submarine cable used for some time was not suitable for the shallow-water sections of CANTAT, and hence the cable shown in Fig. 6.23 was developed. This type of cable has a mild-steel strand inside its inner copper conductor, not to provide strength but to give the cable the same electrical characteristics as the lightweight cable. The outer conductor consists of six copper tapes and is separated from the inner

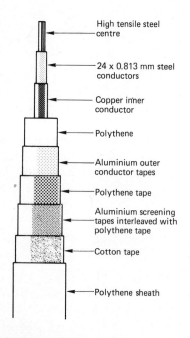

Fig. 6.22 Construction of a lightweight submarine cable

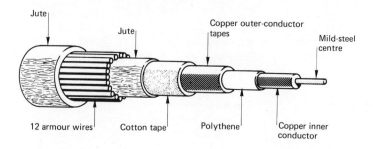

Fig. 6.23 Construction of an armoured submarine cable

conductor by polythene insulation. First cotton tape, impregnated with a corrosion inhibitor, and then a layer of jute are placed around the outer conductor before 12 armour wires are applied for protection. Finally, another layer of jute is applied on top of the armour wires.

Internal Cables

The construction of a cable designed for use within a building which houses telecommunication equipment differs from that of an external cable for several reasons. These are:

(i) Internal cables are generally subjected to considerably more movement and so must be insulated and sheathed by a tough and flexible material.
(ii) Conductors in an internal cable are soldered more often and so the insulating material must be heat resistant.
(iii) Internal cables must be fitted into the space available inside and around equipment, and therefore need to be flexible.
(iv) Since the lengths of internal cable runs are relatively short, small diameter conductors can be employed, typically 0.4 mm and 0.5 mm.

The construction of a 30-pair internal cable is shown by Fig. 6.24. The copper conductors are insulated by p.v.c. and the two conductors forming a pair are twisted together. The first pair, numbered 1 in the figure, is at the centre of the cable and the rest of the pairs are positioned as shown. The cable core is then wrapped with polyethalene-tetraphalate tape before enclosure within a p.v.c. sheath.

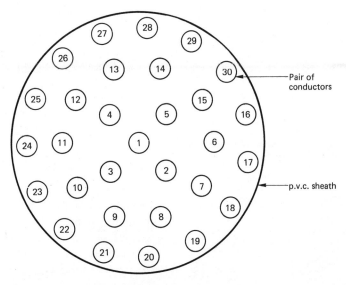

Fig. 6.24 Construction of an internal cable

Loading of Cables

The attenuation of a cable is caused by its conductor I^2R losses and its dielectric V^2G losses. Conductor losses are the greater of the two, and indeed for coaxial cables (except the flexible type) the dielectric losses may be negligibly small. If the current flowing in a line for a given applied voltage could be reduced without a corresponding increase in the line resistance, a reduction in the attenuation could be acheived. At any point along a line the current is equal to the voltage at that point divided by the characteristic impedance of the line; this means that the line current could be reduced by increasing the characteristic impedance. The characteristic impedance of a line depends upon the values of each of the four primary coefficients and could be increased either by increasing the line inductance and/or the line resistance *or* by decreasing the line capacitance and/or leakance. In practice, it is the line inductance which is increased by the insertion of inductors at regular intervals along the length of the line. The practice is known as *loading* the line.

Three loadings are employed in the UK; these are 88 mH every 1.828 km, 120 mH every 1.828 km and 44 mH every 1.81 km, with 88 mH loading being the most common. Fig. 6.25 shows the attenuation/frequency characteristic of a 0.9 mm cable pair with each value of loading inductance. It can be seen that the attenuation of the cable has been reduced at lower frequencies (compare with Fig. 6.14) but rises rapidly at some higher frequency. Clearly the loaded line acts like a low-pass filter.

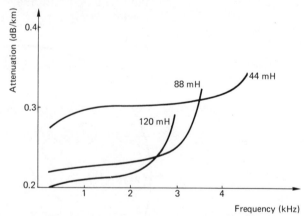

Fig. 6.25 Attenuation/frequency characteristics of loaded lines

The group delay/frequency characteristic of a loaded line depends upon the amount of loading and the length of the line. Fig. 6.26a shows typical group delay/frequency characteristics of loaded lines of 16 km length and having loading inductances of 88 mH/1.828 km, 120 mH/1.828 km and 44 mH/1.81 km. Fig. 6.26b shows group delay/frequency characteristics of different lengths of 88 mH/1.828 km loaded line.

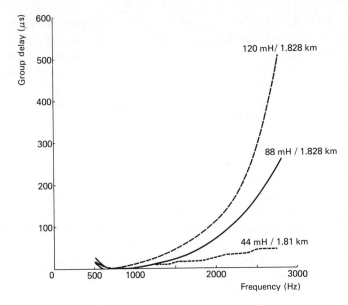

Fig. 6.26a Group delay/frequency characteristics of loaded lines in 10 km lengths with different loadings

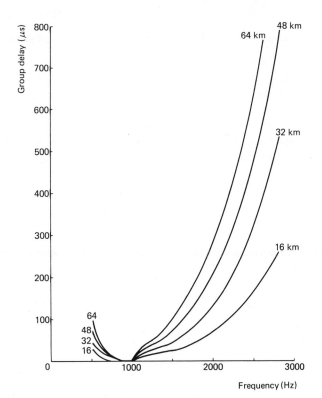

Fig. 6.26b Group delay/frequency characteristics of loaded lines in different lengths of 88 mH/1.828 km loading

The standard loading in the UK telephone network is 88 mH/1.828 km although a number of older cable routes with other values of loading inductance are still in use. As telephone traffic increases, the demand for extra junction circuits intensifies. When spare pairs are available in an existing cable the extra junctions can readily be provided. If, however, spare pairs are not available it will be necessary to install extra circuits. The new junctions can be provided by the installation of another cable but it is generally more economic to convert a junction route to *pulse-code modulation* (p.c.m.) operation. The principles of operation of a p.c.m. system will be considered in Chapter 8; here it will suffice to know that a p.c.m. system provides several telephone channels over a single circuit—at the expense, however, of a much wider bandwidth requirement. The bandwidth of a loaded line is limited to about 4 kHz and this figure is not wide enough to accommodate a p.c.m. system. Because of this, any cables which are to be operated using p.c.m. must first have their loading coils removed.

Use of Transmission Lines as Radio Station Feeders

In radio stations, both for transmission and reception, feeders are necessary to connect the station aerials to the radio transmitters or receivers. At a transmitting station the main requirement of a feeder is to transmit large amounts of power with utmost economy; in a receiving station the primary function of a feeder is to convey the signals picked up by an aerial to the receiver with minimum degradation in signal-to-noise ratio.

Both open-wire and coaxial lines are used as feeders, each having its particular merits. Open-wire feeders are cheaper to provide and to operate than coaxial feeders. The open-wire feeder has a higher characteristic impedance than the coaxial feeder (maximum $Z_0 \simeq 80 \, \Omega$) and is therefore more suitable for feeding the rhombic and log-periodic aerials which are generally used for h.f. point-point radio-telephony links. It is difficult to ensure that the correct spacing is maintained between the conductors of a two-wire feeder, and this may well result in variations from the nominal value of characteristic impedance. Other disadvantages of two-wire feeders are their susceptibility to crosstalk from other feeders, their tendency to radiate at higher frequencies, and their large space requirement. On the other hand, coaxial feeders do not radiate and are not subject to interference from adjacent feeders. Coaxial feeders are also much more convenient than two-wire feeders for use in aerial switching systems, but their use for long distances cannot be economically justified. A compromise solution that has found

TELEPHONE LINES AND CABLES

favour is to use coaxial feeders within the radio station building itself and two-wire feeders from the building to the aerials.

The Effect of Cables on Analogue and Digital Signals

In Chapter 1 the range of frequencies produced by the human voice, by musical instruments and by telegraphy and television systems were considered, and it was seen that the range of frequencies transmitted by a communication system had to be limited for economic and operational reasons.

For a reproduced sound signal to appear as natural as possible within the necessary bandwidth restrictions, it is essential that the original amplitude relationships between the fundamental component and the various harmonics are retained. As a signal is propagated along a transmission line it will be attenuated, and this attenuation is greater at the higher frequencies than at the lower. The effect of line attenuation is, therefore, to reduce the amplitudes of the harmonics relative to the fundamental and make the received sound seem unnatural. The effect on short telephone lines is slight and can be tolerated but longer telephone circuits and all-music circuits are normally *equalized* to overcome this problem. A circuit, known as an EQUALIZER, is fitted to the receiving end of a line and is adjusted to have an attenuation/frequency characteristic which is the inverse of the attenuation response of the line. The total attenuation of the circuit is then the sum of the attenuations of the line and of the equalizer and is more or less constant (see Fig. 6.27). Clearly, the circuit loss at lower frequencies has been increased but this can be countered by the use of a line amplifier.

With telegraphy and television transmissions the signal waveform must be retained and this means that the various component frequencies must keep both their amplitude and phase relationships relative to one another and to the fundamental component. Thus, both amplitude/frequency and group delay/frequency distortion must be small. The bandwidth occupied by a telegraphy signal is so small that distortion is negligible on short circuits while all longer circuits are routed over a multi-channel system. The bandwidth occupied by a television signal is several megahertz and consequently both amplitude and group delay/frequency distortion are of importance. Lines for the transmission of television signals are generally fitted with both attenuation and group-delay equalizers. The purpose of an attenuation equalizer has been mentioned previously (see Fig. 6.27); the function of a group delay equalizer, fitted at the receiving end of a line, is to introduce frequency variable group delay in addition to that provided by

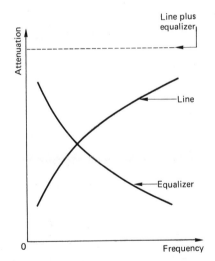

Fig. 6.27 Equalization of a line

the line so that the total group delay is the same at all frequencies.

Digital data signals may also be affected by attenuation and group delay distortions. When frequency shift modulation is used, the effect of line attenuation may be to reduce the maximum possible data transmission rate. The attenuation of the line may be so great at 2100 Hz that the receiving equipment is unable to respond correctly to the low-level incoming signal. Generally, the smaller attenuation at 1700 Hz will permit lower speed transmissions to be received satisfactorily. The effect of group delay/frequency distortion is to delay 2100 Hz pulses to a greater extent than 1300 Hz pulses, with the result that the pulses arrive at incorrect instants in time. The effect is to make the pulses overlap so that binary 1 appears longer and binary 0 shorter than they should. In extreme cases the receiving equipment may not be able to detect a 0 in between two or more 1 pulses and errors will result.

Exercises

6.1. Make careful sketches to show the construction of *two* of the following: (*a*) an aerial feeder for a high-power medium-frequency broadcast transmitter, (*b*) a submarine coaxial cable, (*c*) a star-quad type trunk telephone cable.

Explain the reasons for the choice of insulating materials used, and quote an approximate figure for a typical characteristic impedance for each case you have described.

(C&G)

6.2. Tabulate the relative advantages and disadvantages of a balanced open-wire transmission line and an unbalanced coaxial feeder to connect a short-wave transmitter to its aerial.

(C&G)

6.3. Sketch a multi-tube coaxial cable suitable for inland use. Give typical dimensions and indicate the materials used.

A coaxial pair has a loss of 20 dB at a frequency of 1 MHz. Calculate its loss at 9 MHz, assuming that dielectric loss may be neglected.

(C&G)

6.4. Sketch the arrangement of a 4-tube coaxial cable for inland use. Show typical dimensions and mention the materials used.

A coaxial cable has a loss of 3.9 dB/km at 1.2 MHz. If dielectric loss is negligible, what is the loss of 6 km of cable at 4.34 MHz?

(C&G)

6.5. Sketch in detail the construction of (*a*) a multi-pair star-quad telephone cable, and (*b*) a multi-tube coaxial cable. State the materials used, give typical dimensions and characteristic impedances, and indicate for what types of services these cables are normally used.

(C&G)

6.6. Sketch the arrangement of a 4-tube coaxial cable for inland use. Show typical dimensions and mention the materials used. State the relationship between attenuation and frequency at high frequencies for a coaxial cable whose dielectric loss is negligible.

A broadband transmission system using such a cable occupies the band 1 MHz to 10 MHz. The cable has a loss of 4.21 dB/km at 1 MHz and it is required that the attenuation between adjacent repeaters should not exceed 40 dB. Calculate the maximum repeater spacing. (C&G)

6.7. Define the term *characteristic impedance* as applied to a transmission line. Sketch the construction of a 4-tube coaxial cable suitable for inland telephony use, giving typical dimensions and showing the materials used. Sketch a typical attenuation/frequency characteristic for a coaxial tube. (C&G)

6.8. Sketch the arrangement of a 4-tube coaxial cable for inland use. Show typical dimensions and mention the materials used. Why is each tube lapped with steel tape?

A coaxial cable whose dielectric loss is negligible has an attenuation of 2.9 dB/km at 0.5 MHz. Calculate the frequency at which the loss of a 6 km length will be 54 dB. (C&G)

6.9. Sketch the construction of coaxial cable for use in *either* an inland *or* a submarine carrier telephony system. State the most usual characteristic impedance for each type of cable.

The loss of 6 km of coaxial cable is 4.5 dB at 4.34 MHz. Calculate the loss of 3 km of the cable if the frequency of operation were extended to 12 MHz. Neglect dielectric losses. (C&G)

6.10. What are the advantages and disadvantages of open-wire balanced transmission lines and unbalanced coaxial cables when used to provide a remote connection between a receiver and its aerial?

State the basic differences in providing the end-connections of the following transmission lines to the aerial and the receiver: (a) a balanced two-wire feeder, (b) a coaxial cable. (C&G)

6.11. (a) Describe with a dimensioned sketch, a coaxial cable suitable for connecting a very-high-frequency aerial to a receiver.

(b) Using the dimensions you have selected, calculate the characteristic impedance of the cable.

(c) Sketch a typical attenuation/frequency characteristic for a cable of this type. (C&G)

6.12. With the aid of a sketch showing approximate dimensions, describe the construction of *either* an assembly of four coaxial pairs for inland use, *or* a submarine coaxial-pair cable, both types of cable being suitable for carrying multi-channel telephony transmissions. From the dimensions you have given calculate the characteristic impedance of the cable.

Sketch the variation of loss with frequency and state the normal range of frequencies in common use. (C&G)

6.13. (a) Describe, with the aid of a sketch, the construction of a coaxial cable for use with a submarine cable system.

(b) If the available channel power from a repeater in the system is 3 mW and the required input channel power is 0.1 μW, calculate the required repeater spacing given that the attenuation of the cable for that channel is 1 dB/km.

6.14. (a) Describe, with the aid of a sketch, the construction of a coaxial cable suitable for use in inland multi-channel telephony systems.

(b) A multi-channel telephony system of 4 MHz bandwidth is transmitted through a coaxial cable. Calculate the signal-to-noise ratio in a 4 kHz channel at 4 MHz after transmission through 9 km of cable, given the following: transmitted signal power = 10 mW, attenuation of cable = $2.3\sqrt{f}$ dB/km, noise level in cable = kTB watts, where the absolute temperature $T = 290°$, B is the bandwidth in hertz, f is the frequency in MHz, $k = 1.38 \times 10^{-23}$ J/°. (C&G)

6.15. A generator of e.m.f. 50 V and internal impedance 600 Ω is applied to a line having a characteristic impedance of 600 Ω and an attenuation coefficient of 1 dB/km. Calculate the current flowing in its correctly terminated load resistance if the line is 20 km long.

6.16. A correctly terminated transmission line has $Z_0 = 600$ Ω, $\alpha = 0.8$ dB/km and $\beta = 25°$/km and is fed by a generator of 600 Ω impedance and 10 V e.m.f. The power dissipated in the load is 5 mW. Calculate the length of the line.

6.17. (a) Explain the difference between the phase velocity and the group velocity of a signal.

(b) What is meant by group delay, and by group delay/frequency distortion?

(c) Explain the effects that group delay/frequency distortion can have upon (i) an analogue signal and (ii) a digital signal.

(d) Draw typical group delay/frequency characteristics for (i) an audio cable (ii) a loaded audio cable, (iii) a channel in a multi-channel system.

Short Exercises

6.18. Make a list of the primary and secondary coefficients of a transmission line. Draw graphs to show how each of the primary coefficients varies with frequency.

6.19. Draw the cross-sectional diagram of an internal multi-pair cable. State the materials employed.

6.20. Draw the cross-sectional diagram of a multi-tube coaxial cable. State the materials employed.

6.21. Draw the cross-sectional diagram of an audio frequency star-quad cable. State the materials employed.

6.22. Draw the cross-sectional diagram of a submarine type of coaxial cable. State (i) the materials used and (ii) whether the cable described is suitable for use in shallow or in deep waters.

6.23. Sketch typical attenuation/frequency and group delay/frequency curves for unloaded and loaded audio frequency cables.

6.24. Explain why (i) a cable possesses attenuation and (ii) why this attenuation increases with increase in frequency.

6.25. What is meant by the characteristic impedance of a cable? A transmission line has a characteristic impedance of 750 Ω and is correctly terminated. Determine the input impedance of the line.

6.26. Explain what is meant by the terms *phase velocity* and *group velocity* as applied to a transmission line.

6.27. A signal consisting of a 5 kHz wave and its third harmonic is transmitted over a transmission line. The phase change coefficient of the line is 10°/km at 5 kHz and 24°/km at 15 kHz. Calculate the group velocity of the signal.

7 Two-wire and Four-wire Circuits

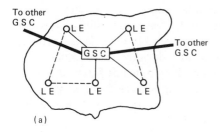

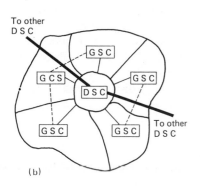

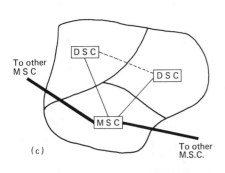

Fig. 7.1 Trunk switching network

The telephone network of Great Britain is divided into local lines, junctions and trunks as explained in Chapter 6. Local lines consist of pairs of wires that connect the individual telephone subscribers to their local telephone exchange (l.e.), junctions are two-wire circuits that may or may not be amplified and connect nearly all telephone exchanges together, and trunks are amplified four-wire circuits that connect distant exchanges together. Long-distance telephone lines are extremely expensive and it is not economically possible to directly conect every exchange in the network to every other exchange; direct trunks are only provided between two exchanges when justified by the traffic carried. The remainder of the trunk traffic is, as would be expected, between exchanges which are not physically located near to one another, and this traffic is routed via trunk switching exchanges known as *group switching centres* (g.s.c.).

A group switching centre also functions as the local telephone exchange for the area in which it is situated (Fig. 7.1*a*). Each group switching centre collects trunk traffic from the local exchanges in its area and has, in turn, trunks to one or more *district switching centres* (d.s.c.). A district switching centre acts as the trunk switching centre for a number of group switching centres (Fig. 7.1*b*). This stage in the switching network is necessary because it is economically impossible to fully interconnect all the group switching centres.

Lastly, a number of district switching centres are chosen to act as a *master switching centre* (m.s.c.) as in Fig. 7.1*c*. Each master switching centre acts as a trunk switching centre for a number of district switching centres. All the master switching centres in the network are fully inter-connected by direct routes.

Whenever justified economically by the telephone traffic, direct routes are provided between two exchanges, for example, between two group switching centres, between a district switching centre and a group switching centre in another district, or between two district switching centres. These direct routes are known as *auxiliary routes* and they supplement the basic network. It has been estimated that about 85% of trunk traffic is routed over the auxiliary network.

Local Lines

The connection between a subscriber and his local telephone exchange consists of a pair of wires in a telephone cable. Since a large telephone exchange may have up to 10 000 subscribers the local line network can be quite complicated, particularly because provision must be made for fluctuating demand. The local line network is provided on the basis of forecasts made of the future demand for telephone service, the object being to provide service on demand and as economically as possible. Since the demand fluctuates considerably there is the problem of forecasting requirements and deciding how much plant should be provided initially and how much at future dates. No matter how carefully the forecasting is carried out, some errors always occur and allowance for this must be made in the planning and provision of cable, i.e. the local line network must be flexible. A network must be laid out so that the situation does not arise where potential subscribers cannot be given service in some parts of the exchange area while in other parts spare cable pairs remain. The modern way of laying out a local line network is shown in Fig. 7.2. Each subscriber's telephone is connected to a distribution point, such as a terminal block on a pole or a wall. The distribution points are connected by small distribution cables to pillars, a pillar being a street structure that provides flexibility because it allows any

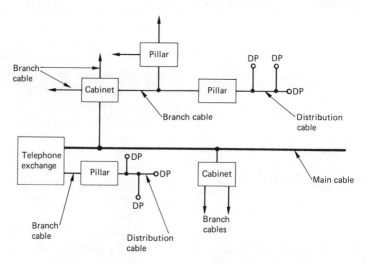

Fig. 7.2 Layout of a telephone exchange area
DP = distribution point

incoming pair to be connected to any outgoing pair. The pillars are connected by larger branch cables to cabinets; these have the same function as pillars but are larger. Finally, main cables connect the cabinets to the telephone exchange.

Junction and Trunk Circuits

Junctions interconnect exchanges that are less than about 10 km apart and trunks link exchanges even further apart. Junctions and the shorter trunks are generally operated using a *two-wire circuit* while the longer trunks are all *four-wire circuits*.

When a signal is transmitted over a telephone line it is attenuated and for all but the shortest circuits will need amplification. Circuits connecting district and/or master switching centres are operated with zero overall loss, group switching centre-to-district switching centre trunks have a loss of 3.5 dB, and group switching centre-to-local exchange junctions have a loss which must not be greater than 4.5 dB. The amplification necessary to achieve these overall loss figures is provided at a number of points along the length of a line to ensure that the signal-to-noise ratio is maintained well above the minimum acceptable figure.

Two-wire Circuits

A two-wire circuit is one that is operated over a single pair of conductors in a telephone cable with signals passing in both directions at the same time and in the same frequency band. When amplifiers of conventional type are employed, the circuit must be split into two separate parts at each amplifier (or repeater) station, each handling one direction of transmission since amplifiers are essentially unidirectional devices.

Fig. 7.3 shows a two-wire circuit that has amplification provided at three separate points along its length. At each amplification point the signal path is divided into two by *hybrid*

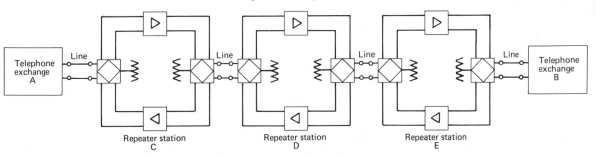

Fig. 7.3 A two-wire amplified circuit

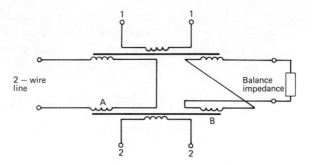

Fig. 7.4 A terminating set

transformers or *terminating units*. A hybrid coil or terminating set (Fig. 7.4) is a device used to convert a 2-wire circuit into a 4-wire circuit. The operation of the unit is as follows: assume that, as is usually the case, the impedances of the circuits connected across terminals 1–1 and 2–2 are equal. A signal applied across the 2-wire terminals of the unit will cause a current to flow and this current will induce an e.m.f. into each of the windings connected across terminals 1–1 and 2–2. The currents in these circuits are of equal magnitude and so they induce equal e.m.f.s into the balance circuit. The current flowing in the balance impedance is therefore zero. The power contained in the signal applied to the 2-wire terminals is divided equally between terminals 1–1 and 2–2 so that the loss between these terminals and the input is 3 dB. In addition, transformer losses of about 1 dB are also present.

When a signal is applied to the terminals 2–2 of the circuit, the current which flows induces e.m.f.s, with the same polarity, into both winding A and winding B. Currents then flow in the 2-wire and balance circuits that induce e.m.f.s of opposite polarity into the winding connected across terminals 1–1. If the balance impedance is adjusted to be equal to the impedance of the 2-wire line, these induced e.m.f.s will be equal and they will cancel. Zero current will then flow at terminals 1–1. The signal power applied to terminals 2–2 is divided between the 2-wire and balance circuits, giving a total loss of 4 dB between terminals 2–2 and the 2-wire terminals. The loss between the terminals 2–2 and 1–1 is very high since little, if any, current flows at terminals 1–1.

An alternative circuit for a terminating unit is shown in Fig. 7.5. A description of its operation will be left as an exercise for the reader (see Example 7.10).

The operation of the terminating unit depends upon the accuracy with which the balance impedance simulates the impedance/frequency characteristic of the two-wire line. If the balance impedance does not exactly simulate the line impedance at all frequencies, the loss of the terminating unit between the terminals 2–2 and 1–1 will be reduced and some

TWO-WIRE AND FOUR-WIRE CIRCUITS   117

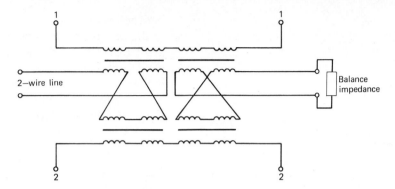

Fig. 7.5 Another terminating set

energy will pass from the output of one amplifier into the input of the other. This fed-back energy will be amplified and some of it will appear at the input of the first amplifier and so on. If the circuit is not to oscillate, that is if it is to be *stable*, the loop losses of the circuit must be greater than the loop gains. Since accurate matching between two-wire lines and balance networks at each amplifying point is difficult to achieve in practice, the probability that a two-wire circuit will be unstable increases with increase in the number of amplifiers. It is customary, therefore, to restrict the use of amplified two-wire circuits to lines of such length that amplification at only one point is sufficient.

Amplification may be provided in a two-wire circuit, if the line loss is not greater than 11 dB, by a two-wire repeater inserted as near to the middle of the circuit as possible. The repeater may incorporate conventional amplifiers (Fig. 7.6) or may be a negative-impedance amplifier (Fig. 7.7). The former

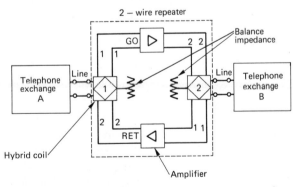

Fig. 7.6 The two-wire amplifier

type consists of two amplifiers—valve or transistor—and two hybrid coils connected as shown.

Speech signals reaching the repeater from exchange A divide equally at hybrid coil 1, and the signals arriving at terminals 1,1 are 4 dB below the level at the 2-wire terminals, and are amplified by the GO amplifier before application to terminals 2,2 of hybrid coil 2. The overall gain of the repeater between the 2-wire terminals of the two hybrid coils is equal to the gain of the amplifier minus 8 dB. Typically, the amplifier gain might be 16 dB when an input level of, say, −3 dBm would produce an output level of +5 dBm. The signals appearing at terminals 2,2 of hybrid coil 1 are dissipated in the output impedance of the RET amplifier and serve no useful purpose. Since the loss of a hybrid coil between terminals 2,2 and 1,1 is very high, very little energy is fed around the circuit, a necessary condition if the repeater is not to oscillate.

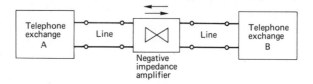

Fig. 7.7 The negative impedance amplifier

A negative-impedance amplifier (Fig. 7.7) is a transistor circuit that provides amplification of signals in both directions by effectively lowering the line resistance. Such an amplifier has a gain variable between 2.5 and 12 dB and is designed for use in conjunction with a particular type of audio telephone cable—naturally the most commonly employed.

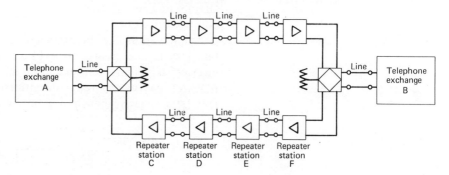

Fig. 7.8 A four-wire amplified circuit

All longer trunks are worked four-wire and the basic arrangement of a four-wire circuit is shown in Fig. 7.8. The lines connecting the telephone exchanges to the terminal repeater stations are operated two-wire but at these stations the circuit is split into GO and RETURN paths. Only two terminating units are used, one at each end of the circuit. There is thus only one possible loop path and the possibility of instability is greatly reduced. Also the risk of instability is not increased by increasing the length of the circuit and/or the number of amplifiers. A four-wire circuit is set up to be stable when the two-wire terminals of the two terminating sets are open-circuited. This practice results in the requirements of the two-wire balance being much less exacting. In the majority of cases a 600 ohm resistor is found to give adequate balance. It is usual to set up a four-wire circuit to have an overall loss of between 0 dB and 4.5 dB so that with a 0 dBm test tone (usually 800 Hz) applied to the two-wire terminals of a terminating set, the amplifier output levels, for that direction of transmission, are all +10 dBm. The gain of the amplifiers used is nominally 27 dB and so the minimum input level to an amplifier is −17 dBm.

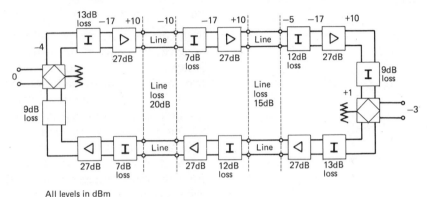

Fig. 7.9 Typical losses and gains for a four-wire circuit

In practice, the level is very often greater than this and the amplifier must then be preceded by an attenuator of suitable loss. Fig. 7.9 shows a four-wire circuit of 3 dB overall loss and gives the levels to be expected at different points in the circuit. The line losses quoted would be at the maximum frequency transmitted.

So far the four-wire circuit has been taken as consisting of two pairs in some telephone cables but, in practice, a four-wire circuit may be routed, wholly or partly, over a single channel in a multi-channel telephony system (see Chapter 8).

A comparison of the relative merits of two-wire and four-wire circuits shows that while the two-wire circuit is economical in the use of cable pairs—an important consideration—the four-wire circuit requires fewer amplifiers for a given length of line and is *much* easier to set up and maintain.

The Interconnection of Junction and Trunk Circuits

The need often arises in a telephone network for trunk and/or junction circuits to be connected in tandem in order to route a call from one exchange to another. Short two-wire circuits are easily switched by manual or automatic means but the problem is somewhat more difficult for four-wire circuits. The automatic equipment required to switch four-wire circuits directly is

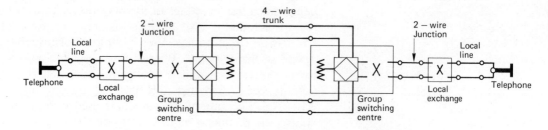

Fig. 7.10 Two-wire switching of trunk circuits

both complex and costly, and for many years it has been the practice to convert four-wire circuits to two-wire before switching. Fig. 7.10 shows a typical connection that involves two local exchanges and two group switching centres. The points marked X denote switching points in telephone exchanges; such a point may consist of automatic equipment or a telephone switchboard. For s.t.d. calls, of course, all the switching points are automatic.

The two-wire switching of four-wire trunk circuits suffers from the disadvantages that (a) there is a 4 dB loss through each terminating set and (b) the overall loss of each trunk circuit cannot be less than about 1.5 dB because of the instability problem and this means that the switching losses cannot be compensated for. These disadvantages naturally assume greater importance as the number of switching points is increased, and therefore all switching in district and master

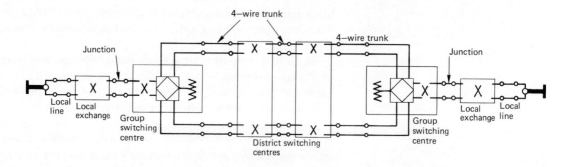

Fig. 7.11 Four-wire switching of trunk circuits

switching centres is carried out on a four-wire basis. The arrangement of a typical four-wire switched connection is shown in Fig. 7.11 in which switching points are again marked X. It can be seen that the connection is four-wire between the two terminal group switching centres and that the group switching centre to local exchange link is worked two-wire. Any losses introduced at the four-wire switching points can be compensated for by increasing the amplifier gains.

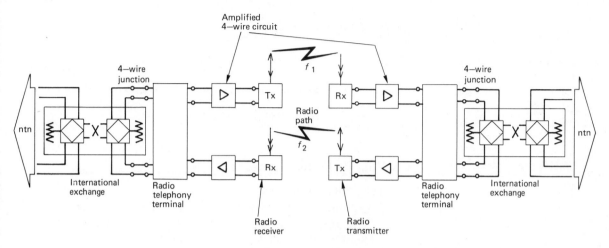

Fig. 7.12 An international telephony circuit (ntn = national telephone network)

International Circuits

Many international circuits are routed on multi-channel carrier-telephony circuits passing over submarine cable, radio link, or earth satellite and these will be discussed later. Here the intention is to mention the method used to interconnect two subscribers in different countries by a four-wire radio link operating in the high-frequency band 3–30 MHz. Fig. 7.12 shows a typical arrangement.

A form of s.s.b. working, known as independent sideband (i.s.b.), enables four commercial-quality 25–3000 Hz bandwidth speech circuits to be accommodated in a 12 kHz bandwidth. A subscriber in one country is connected, via the junction and trunk network of that country, to his international telephone exchange. Here his call is established via a radio circuit to the required country and the telephone network in that country. Signals passing between the two subscribers are at audio frequency up to the radio link itself; at the transmitter the signals amplitude-modulate a carrier in the h.f. band and the resulting waveform is radiated to the distant receiver. Different frequencies are used for the two directions of transmission over the radio link to eliminate the possibility of singing around the loop.

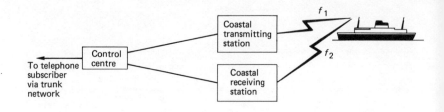

Fig. 7.13 Shore-to-ship telephone connection

Ship Radio-telephones

Telephonic communication is often required between a telephone subscriber and a ship at sea and the arrangement for setting up such a connection is shown in Fig. 7.13. The telephone subscriber is connected, via the trunk network, with the control centre. The control centre is linked by cable to a number of coastal transmitting and receiving radio stations, each of which transmits to or receives from a different part of the world. The control centre establishes the required connection with a distant ship via the appropriate pair of radio stations. To provide good coverage of a particular area of the world each transmitting station transmits on several different frequencies at the same time. The actual frequencies depend upon the part of the world concerned and are given in Chapter 2.

Exercises

7.1. By reference to a block schematic diagram, explain the equipment needed to provide a repeatered audio junction circuit on (*a*) a 2-wire basis, and (*b*) a 4-wire basis. Quote typical losses for each part of the circuit.
State the relative advantages of 2-wire and 4-wire operation. (C&G)

7.2. (*a*) Draw a labelled block diagram of a simple communication system in which a 2-wire to 4-wire termination is used. (*b*) State the purpose of the 2-wire to 4-wire termination. (*c*) Draw the circuit of a 2-wire to 4-wire termination and briefly outline its operation. (*d*) State the order of loss in decibels that you would expect in the 2-wire to 4-wire termination. (C&G)

7.3. (*a*) With the aid of a diagram outline the operation of a 2-wire to 4-wire terminating set in an audio-frequency telephone circuit. (*b*) Draw a labelled block diagram to illustrate the use of a 2-wire to 4-wire terminating set in *either* a simple line system *or* a simple radio system. (C&G)

7.4. (*a*) Draw suitably labelled block diagrams to show the equipment and links for a repeatered audio junction circuit on (i) a 2-wire basis, (ii) a 4-wire basis. (*b*) Briefly explain the purpose of the terminating sets used in each case. (*c*) State the relative advantages of 2-wire and 4-wire operation. (C&G)

7.5. A four-wire circuit is to be set up between two towns A and B with amplification provided at both ends of the circuit and at one intermediate town C. The loss of the cable between towns A and C is 22 dB and the cable loss between towns C and B is 18 dB. Draw a block diagram of the circuit assuming (i) an overall loss of −4.5 dB, (ii) that standard line amplifiers are used having a gain of 27 dB and an output level of +10 dBm.

7.6. Repeat Exercise 7.5 assuming the cable loss between towns A and C to be 25 dB and between towns B and C to be 21 dB.

7.7. Draw, and explain, a block diagram to show how an international radio-telephony link is integrated with the national telephone networks of two different countries.

7.8. (a) Draw block diagrams to show the equipments and connections for the following systems: (i) an amplified audio junction circuit between two telephone exchanges, (ii) a public ship-to-shore long-distance telephone service.

(b) Briefly explain the purpose of the terminating sets used in either (i) or (ii) above.

(c) Draw the circuit diagram of a typical terminating set.

(C&G)

7.9. Discuss how a 4-wire circuit could become unstable. Is this more, or less, likely to occur with a radio circuit than with a line circuit? Give reasons for your answer.

7.10. Explain the operation of the terminating unit whose circuit is given in Fig. 7.5.

Short Exercises

7.11. What are the function of (i) a group switching centre and (ii) a district switching centre in a trunk network?

7.12. Draw the circuit diagram of a terminating unit and describe its operation.

7.13. What are the relative merits of 2-wire and 4-wire operation of amplified line circuits?

7.14. Explain why there is often a need for separate paths for transmission and reception in amplified line circuits.

7.15. Draw a block diagram of a 4-wire amplified line circuit and explain the paths taken by the signals transmitted in each direction.

8 Frequency-division and Time-division Systems

A telephone network connects customers with telephone, telegraph, or data system equipment to any other customer with whom a link is required. When the number of customers is fairly small and they are physically located in the same neighbourhood, it is economically feasible to provide a communication network which consists entirely of physical cable pairs. For example, the telephone subscriber is connected to his local telephone exchange by a pair in the local line network, and can be switched by the exchange equipment to other subscribers connected to that exchange or to a junction to another exchange. The term *space division* is often employed to describe such systems which require a separate cable pair for each circuit, that is systems which do not depend on either frequency-division or time-division. Telephone cable is extremely expensive and, together with the associated ductwork, accounts for the major part of the cost of providing a communication link between two points. It is therefore desirable to increase the traffic-carrying capacity of trunk circuits (in particular) and junctions by the use of either frequency-division multiplex (f.d.m.) or time-division multiplex (t.d.m.) systems. The basic principles of both t.d.m. and f.d.m. have been outlined in Chapter 2; in this chapter the basic principles of multi-channel f.d.m. carrier and coaxial telephony systems and of pulse-code modulation systems will be considered.

Balanced Modulators

Frequency-division multiplex telephony systems utilise single sideband suppressed carrier amplitude modulation of a number of carriers of appropriate frequency. The unwanted sideband is suppressed during the modulation process by means of a *balanced modulator*.

Many balanced modulators, particularly those in multichannel line systems, do not utilize the square-law characteristics of a diode or transistor but, instead, use the device as an electronic switch. When a transistor or diode is forward biased its resistance is low, and when it is reverse biased its resistance is high. Provided the carrier voltage is considerably greater than the modulating signal voltage, the carrier will control the switching of the device. Ideally, a device should have zero forward resistance and infinite reverse impedance, and this will be assumed in the circuits that follow.

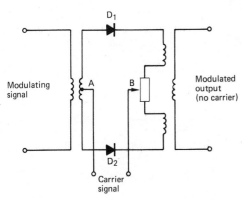

Fig. 8.1 Single balanced modulator

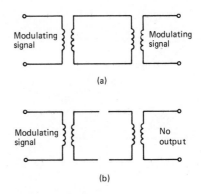

Fig. 8.2 Operation of a single balanced modulator

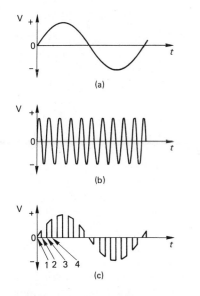

Fig. 8.3 Output waveform of a single balanced modulator
(a) Modulating signal
(b) Carrier wave
(c) Output waveform

Fig. 8.1 shows the circuit of a single-balanced diode modulator. During the half-cycles of the carrier waveform that make point A positive with respect to point B, diodes D_1 and D_2 are forward biased and have zero resistance. The modulator may then be redrawn as shown in Fig. 8.2a; obviously the modulating signal will appear at the output terminals of the circuit. Similarly, when point B is taken positive relative to point A, the diodes are reverse biased and Fig. 8.2b represents the modulator. The action of the modulator is to switch the modulating signal on and off at the output terminals of the circuit. The output waveform of the modulator can be deduced by considering the modulating signal and carrier waveforms at different instants. Consider Fig. 8.3: during the first positive half-cycle of the carrier wave a part of the modulating signal appears at the output (1–2 in Fig. 8.3c); in the following negative half-cycle the modulating signal is cut off (2–3); in the next positive half-cycle the corresponding part of the modulating signal again appears at the output terminals (3–4); and so on.

Analysis of the output waveform shows that it contains the upper and lower sidefrequencies of the carrier ($f_c \pm f_m$), the modulating signal f_m, and a number of higher, unwanted frequencies, but the carrier component is *not* present. In practice, of course, diodes are non-ideal and this has the effect of generating further unwanted frequencies and of reducing the amplitude of the wanted sidefrequency. Some carrier leak also occurs, and a potentiometer is often included to adjust for minimum leak.

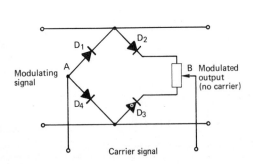

Fig. 8.4 Cowan modulator

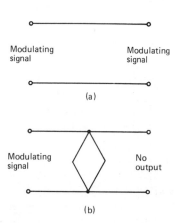

Fig. 8.5 Operation of the Cowan modulator

Another circuit that performs the same function is the *Cowan modulator* (Fig. 8.4). The carrier voltage is applied across points A and B and switches the four diodes rapidly between their conducting and non-conducting states. When point B is positive with respect to point A all four diodes are reverse biased and the modulator may be represented by Fig. 8.5a; during the alternate carrier half-cycles Fig. 8.5b applies. The modulator output therefore consists of the modulating signal switched on and off at the carrier frequency. The output waveform is the same as that of the previous circuit (Fig. 8.3) and contains the same frequency components. The Cowan modulator, however, does not require centre-tapped transformers and it is therefore cheaper; it also possesses a self-limiting characteristic (i.e. the sidefrequency output voltage is proportional to the input signal level only up to a certain value and thereafter remains more or less constant).

Sometimes it is necessary to suppress the modulating signal as well as the carrier wave during the modulation process, and then a *double balanced modulator* is used. Fig. 8.6 shows the circuit of a *ring modulator*. During half-cycles of the carrier wave when point A is positive relative to point B, diodes D_1 and D_2 are conducting and diodes D_3 and D_4 are not; D_1 and

FREQUENCY-DIVISION AND TIME-DIVISION MULTIPLEX SYSTEMS

D_2 therefore have zero resistance and D_3 and D_4 has infinite resistance; Fig. 8.7a applies. Whenever point B is positive with respect to point A, diodes D_1 and D_2 are non-conducting, D_3 and D_4 are conducting, and Fig. 8.7b represents the modulator. It is evident that the direction of the modulating signal current at the modulator output terminals is continually reversed at the carrier frequency.

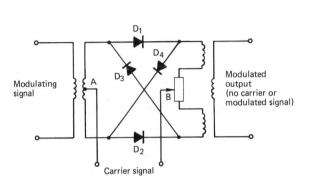

Fig. 8.6 Ring modulator

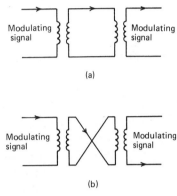

Fig. 8.7 Operation of the ring modulator

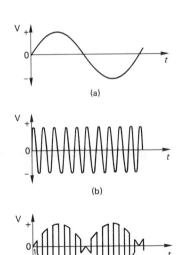

Fig. 8.8 Output waveform of a ring modulator
(a) Modulating signal
(b) Carrier wave
(c) Output waveform

The output waveform of a ring modulator is shown in Fig. 8.8c and can be deduced from Figs. 8.8a, b. Whenever the carrier voltage is positive the modulating signal appears at the modulator output with the same polarity as (a) (see points 1–2 and 3–4 at (c)). Whenever the carrier voltage is negative the polarity of the modulating signal is reversed (points 2–3 and 4–5).

Analysis of the output waveform shows the presence of components at the upper and lower sidefrequencies of the carrier wave and a number of higher, unwanted frequencies. Both the carrier wave *and* the modulating signal are suppressed.

Frequency-Division Multiplex Systems

The concept of f.d.m. was discussed in Chapter 2; here the principles of operation of carrier telephony and coaxial systems will be considered.

The 12-Channel CCITT Carrier Group

Most of the coaxial telephony systems in use in Great Britain consist of a suitable combination of a number of CCITT

128 FREQUENCY-DIVISION AND TIME-DIVISION MULTIPLEX SYSTEMS

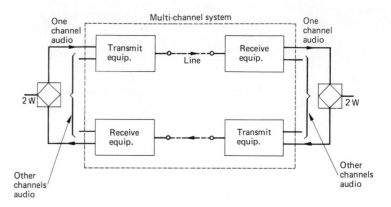

Fig. 8.9 Principle of a multi-channel telephony system

12-channel carrier groups. Circuits routed over a multi-channel system are operated on a 4-wire basis as shown in Fig. 8.9. Fig. 8.10a is a block schematic of the transmitting equipment required for channels 1 and 2 of the 12-channel group. The audio input signal to a channel is applied to a balanced modulator of the Cowan type together with the carrier frequency appropriate to that channel. The input attenuator ensures that the carrier voltage is 14 dB higher than the modulating signal voltage, as required for correct operation of the modulator. The output of the modulator consists of the upper and lower sideband products of the modulation process together with a number of unwanted components.

Following the modulator is another attenuator whose purpose is twofold; firstly, it ensures that the following band-pass filter is fed from a constant-impedance source—a necessary condition for optimum filter performance—and secondly, it enables the channel output level to be adjusted to the same value as that of each of the other channels. The filter selects the lower sideband component of the modulator output, suppressing all other components. To obtain the required selectivity, channel filters utilizing piezoelectric crystals are employed. The outputs of all the twelve channels are combined and fed to the output terminals of the group. The channel carrier frequencies, specified by the CCITT, are listed in Table 8.1. The table also gives details of the passband of each channel filter; it should be noted that the bandwidths correspond to an audio bandwidth of 300–3400 Hz. (Table 8.1 is on p. 136.)

The transmitted bandwidth is therefore 60.6–107.7 kHz, or approximately 60–108 kHz.

The equipment appropriate to channels 1 and 2 at the receiving end of the CCITT 12-channel group is shown in Fig. 8.10b. The composite signal received from the line, occupying the band 60–108 kHz is applied to the twelve, paralleled, channel filters. Each filter selects the band of frequencies

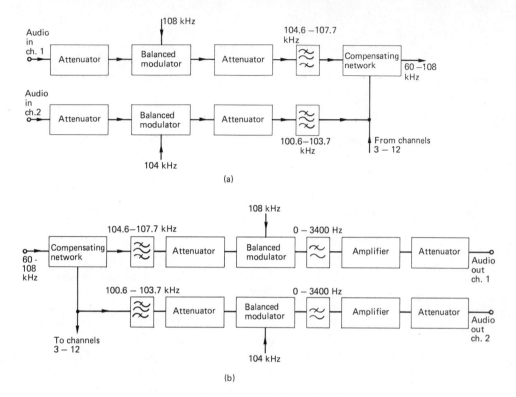

Fig. 8.10 The CCITT 12-channel telephony group

appropriate to its channel, 104.6–107.7 kHz for channel 1, and passes it to the channel demodulator. The attenuator between the filter and the demodulator ensures that the filter works into a load of constant impedance. The demodulator, a Cowan balanced modulator, is supplied with the same carrier frequency as that suppressed in the transmitting equipment. The lower sideband output of the demodulator is the required audio-frequency band of 300–3400 Hz and is selected by the low-pass filter. The audio signal is then amplified and its level adjusted by means of the output attenuator.

The assembly of the basic 12-channel carrier group can be illustrated by means of a frequency spectrum diagram. The spectrum diagram of a single channel is given in Fig. 8.11a; the actual speech bandwidth provided is 300–3400 Hz but a 0–4000 Hz bandwidth must be allocated per channel to allow a 900 Hz spacing between each channel for filter selectivity to build up. Fig. 8.11b shows the frequency spectrum diagram for the 12 channels forming a group; the carrier frequency of each channel is given and so are the maximum and the minimum frequencies transmitted. It can be seen that all the channels

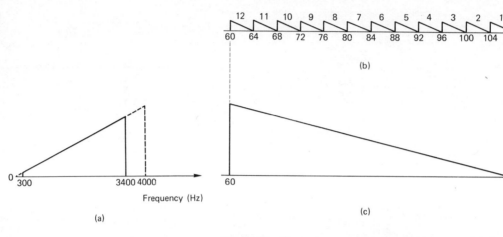

Fig. 8.11 Frequency spectrum diagrams of (*a*) a commercial quality speech channel, (*b*) and (*c*) a CCITT 12-channel group

are *inverted*; that is, the lowest frequency in each channel corresponds to the highest frequency in its associated audio channel, and vice versa. Since all the channels are inverted the group may be represented by a single triangle as shown by Fig. 8.11*c*. The block marked CTE represents the *channel translating equipment*.

The 12-channel system can more conveniently be represented in block diagrams using the system shown in Fig. 8.12. The 12-channel group can be used as a building block for the next larger assembly stage or as a system which can be transmitted to line in its own right. This is also true for the larger assemblies that will be described later in this chapter.

Five 12-channel groups can be combined to form a 60-channel *supergroup* (Fig. 8.13). The block marked GTE represents the *group translating equipment*. Each of the five groups is used to modulate a different carrier frequency, the particular frequencies being marked on the frequency spectrum diagram of Fig. 8.14*a*. Group 1, occupying the frequency band 60–108 kHz, modulates a 420 kHz carrier and the lower sideband of 420–(60–108 kHz) or 312–360 kHz is selected. Group 2 modulates a 468 kHz carrier to produce a lower sideband of 360–408 kHz and so on for the remaining three groups. Since all the groups are erect, the supergroup is erect also as shown by the diagram of Fig. 8.14*b*.

The *hypergroup* consists of 15 supergroups assembled in the manner illustrated by Fig. 8.15, in which the *supergroup translating equipment* is the block marked STE. The frequency spectrum diagram of a hypergroup is given in Fig. 8.16, and shows the individual supergroups making up the hypergroup.

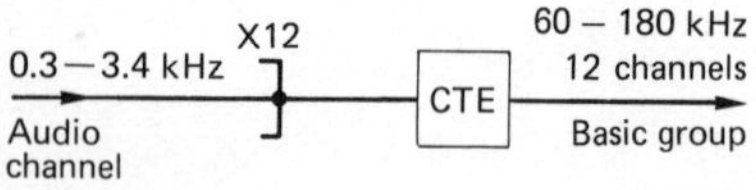

Fig. 8.12 Representation of the 12-channel group

Fig. 8.13 Formation of a super-group

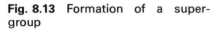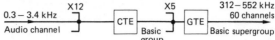

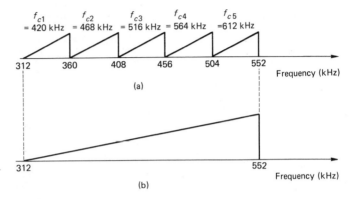

Fig. 8.14 Frequency spectrum diagrams of a supergroup

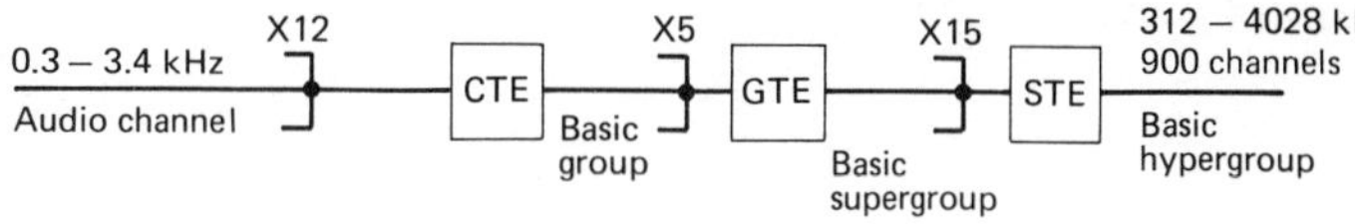

Fig. 8.15 Assembly of a hypergroup

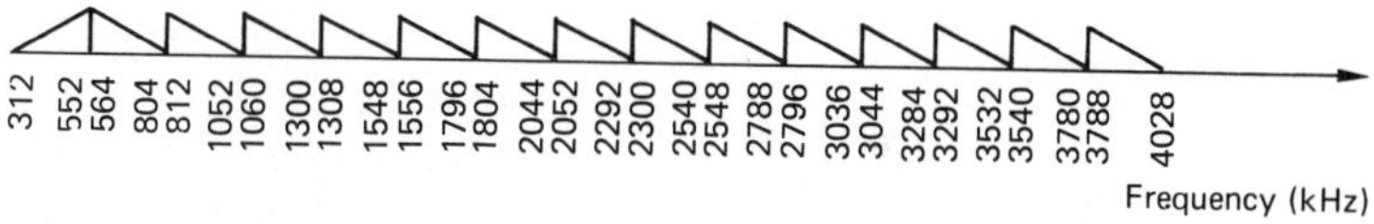

Fig. 8.16 Hypergroup frequency spectrum diagram

One supergroup is left in the supergroup band of 312–552 kHz and is erect. The other 14 supergroups modulate the appropriate carrier frequencies to position the supergroups in the required parts of the frequency spectrum. These 14 supergroups are all inverted.

A hypergroup can be transmitted to line as an 900-channel system or it may be combined with other hypergroups to produce a system with even larger capacity. For example, a 2700-channel system can be produced by combining three hypergroups in the band 312–12 388 kHz. A modern system combines 12 hypergroups together to assemble a 10 800-channel system occupying the frequency band of 4404–59 580 kHz.

An alternative method of combining supergroups is available and is used in some other countries; it is to assemble five supergroups to form a 300-channel *mastergroup*. The five supergroups modulate, respectively, carrier frequencies of 1364 kHz, 1612 kHz, 1860 kHz, 2108 kHz and 2356 kHz to produce the frequency spectrum diagrams shown in Figs. 8.17a and b. The mastergroup can be transmitted to line or it can be used as the building block for a *supermastergroup*. A supermastergroup is an assembly of three mastergroups and provides 900 channels in the frequency band 8516–12 388 kHz. When required, two or more supermastergroups can be combined to form even larger capacity systems.

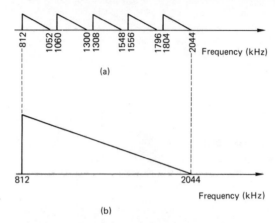

Fig. 8.17 Frequency spectrum diagram of a mastergroup

Time-division Multiplex Systems

The basic principles of a time-division multiplex (t.d.m.) system were outlined in Chapter 2, where it is shown how t.d.m. allows a number of different channels to have access to the common transmission path for a short period of time. The methods of pulse modulation outlined in that chapter are, in practice, rarely employed since a much better performance can be achieved by the use of *pulse code modulation* (p.c.m.).

Pulse Code Modulation

In a p.c.m. system the analogue signal is sampled at regular intervals. The total amplitude range that the signal may occupy is divided into a number of levels each of which is allocated a number. This is shown by Fig. 8.18 in which only 8 sampling levels are shown for clarity but in practical systems many more levels are used. Each time the signal is sampled its instantaneous amplitude is rounded off to the nearest sampling level. The number of this sampling level is then converted into the equivalent binary form and transmitted to line by the appropriate pulse train. Generally, a pulse is sent to represent binary 1 and no pulse is transmitted to indicate binary 0.

The process of approximating the sampled signal amplitudes to the nearest sampling level is known as *quantization*.

FREQUENCY-DIVISION AND TIME-DIVISION MULTIPLEX SYSTEMS   133

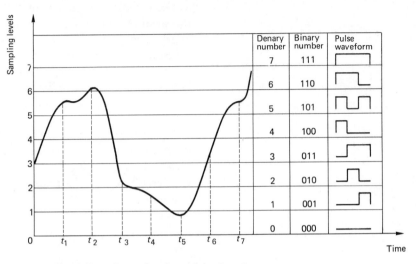

Fig. 8.18 Quantization of a signal

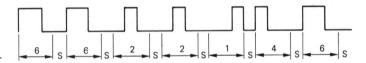

Fig. 8.19 Binary pulse train representing the signal shown in Fig. 8.18

Consider Fig. 8.18. The signal waveform is sampled at time instants t_1, t_2, t_3, etc. At time t_1 the instantaneous signal amplitude is in between levels 5 and 6 but since it is nearer to level 6 it is approximated to this level. At instant t_2 the signal voltage is slightly greater than level 6 but is again rounded off to that value. Similarly the sample taken at t_3 is represented by level 2, the t_4 sample by level 2, the t_5 sample by level 1, and so on. The binary pulse train which would be transmitted to represent this signal is shown in Fig. 8.19. A space, equal in time duration to one binary pulse, has been left in between each binary number in which synchronization information can be transmitted.

A p.c.m. system transmits signal information in *digital form*. The quantization process will result in some error at the receiving end of the system when the analogue signal is reconstituted. The error appears in the form of *quantization noise* and can be reduced only by increasing the number of sampling levels. Unfortunately, this will increase the number of binary digits required to signal the sampling level numbers and this, in turn, means that the bandwidth which must be provided will be wider. For the receiving equipment to be able to decode the incoming binary pulse trains it is only necessary for it to be able to determine whether or not a pulse is present at each instant in time. The characteristics of the pulses, such as their height or their width, are not of importance. Should the attenuation of the transmission path be large enough to degrade the shapes of the pulses to such an extent that errors in reconstituting the signal at the receiver are likely, the pulses can be regenerated (not amplified) at regular intervals along the length of the path. The spacing between the regenerators can be made sufficiently small to ensure that the probability of errors is kept at a very low level. The signal-to-noise ratio at the output of the system is then independent of the length of the transmission path; this is in direct contrast to an analogue system in which the signal-to-noise ratio must always fall as the length of the transmission path is increased. The block schematic diagram of one direction of transmission of a p.c.m. system is shown in Fig. 8.20.

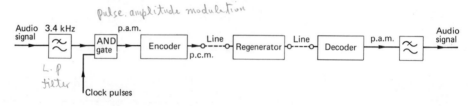

Fig. 8.20 A p.c.m. system

The audio signal is passed through a low-pass filter, to restrict its upper frequency to 3400 Hz, and is then applied to one of the inputs of a two-input AND gate. A rectangular pulse train, known as the *clock* pulse, is applied to the other input terminal. The gate will produce an output only when both the audio signal and a clock pulse are simultaneously present and so the output is a pulse amplitude modulated (p.a.m.) waveform. The p.a.m. waveform is then applied to the encoder whose function is to convert it into the corresponding quantized p.c.m. waveform. The p.c.m. waveform is transmitted along the line, being regenerated at regular intervals as necessary, and applied to the receiving equipment. Here it is first processed by the decoder; the function of this circuit is the inverse of the encoders', that is the decoder converts the incoming p.c.m. waveform into the corresponding p.a.m. signal plus quantization noise. Finally, the p.a.m. signal is passed through the low-pass filter to obtain the required audio signal.

The minimum sampling frequency, i.e. the number of samples taken per second, is equal to twice the highest frequency contained in the analogue signal. Thus, if the highest modulating signal is limited to 3400 Hz by an input low-pass filter, the minimum sampling frequency will be 6800 Hz. In practice, a sampling frequency somewhat greater than the minimum allowable would be used. For example, the BPO p.c.m. tele-

FREQUENCY-DIVISION AND TIME-DIVISION MULTIPLEX SYSTEMS

phony system uses a 8000 Hz sampling frequency for a maximum audio frequency of 3400 Hz. The same system also employs 128 sampling levels which requires 7 binary pulses plus one synchronization pulse per sample. The pulses are 0.625 μs wide and hence each sample effectively occupies a time period of 8×0.625 or 5 μs. A sample is taken every 1/8000 sec or 125 μs and this means that the larger part of each sample period is unoccupied (see Fig. 8.21). The unoccupied part of each 125 μs time period can be used to carry further p.c.m. signals representing the analogue information carried by other channels. Each of the other channels can be applied in turn to the common transmission path by a suitable choice of the clock pulse trains applied to the individual channel gates. The t.d.m. signal is assembled by interleaving the digits proper to each channel, i.e. the signal contains one digit from each channel in turn.

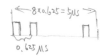
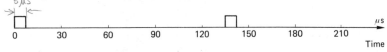

Fig. 8.21 Showing the time that a single channel p.c.m. system occupies a transmission path

The BPO employs a 24-channel system in the junction network which uses an 8 kHz sampling frequency and 128 sampling levels but this system is to be superseded by a CCITT recommended 30-channel system. The BBC use p.c.m. to distribute v.h.f. frequency-modulated stereo signals between studios and radio transmitters. A 32 kHz sampling frequency is used together with 8192 sampling levels.

Relative Merits of f.d.m. and t.d.m.

The advantages of t.d.m., assuming the use of p.c.m., are

(1) The use of regenerators allows an almost distortion-free path regardless of distance, whereas in an f.d.m. system the signal-to-noise ratio gets progressively worse with increase in distance.

(2) Channel selection is achieved by the use of relatively cheap electronic gates and low-pass filters rather than with expensive crystal filters.

(3) The level and phase of the received signal do not depend upon the stability of gain and/or phase shift of the circuits it has been transmitted through.

(4) In an f.d.m. system, care must be taken to avoid non-linearity in amplifiers, etc., because this will result in intermodulation and crosstalk between channels. This is not so for a t.d.m. p.c.m. system.

The main disadvantage of p.c.m. t.d.m. is that a greater bandwidth is needed to transmit a given number of channels than required for the corresponding f.d.m. system. If the signal-to-noise ratio on a p.c.m. transmission path falls below 21 dB, the received signal will deteriorate rapidly and become worse than the corresponding f.d.m. signal.

Table 8.1

Channel no.	Carrier frequency (kHz)	Channel filter passband (kHz)
1	108	104.6–107.7
2	104	100.6–103.7
3	100	96.6– 99.7
4	96	92.6– 95.7
5	92	88.6– 91.7
6	88	84.6– 87.7
7	84	80.6– 83.7
8	80	76.6– 79.7
9	76	72.6– 75.7
10	72	68.6– 71.7
11	68	64.6– 67.7
12	64	60.6– 63.7

Exercises

8.1. A 10 800-channel coaxial system can be assembled using mastergroup and supermastergroup stages. Draw a block schematic diagram of the equipment.

8.2. (a) Describe, with the aid of a circuit diagram and explanatory waveforms, the operation of a ring modulator. (b) Sketch the sideband spectrum obtained. (c) Explain, with the aid of a block diagram, how the modulator described is used in a line transmission circuit to provide one channel in a group of 12. (d) Why is one sideband of the modulated signal normally removed before assembly with the other channels? (C&G)

8.3. (a) State three advantages of using single-sideband compared with double-sideband operation. (b) Draw a block diagram showing the assembly of a 12-channel group of telephone channels. (c) Indicate the type of modulator used and explain why single sideband operation is normally employed. (d) How is the required sideband selected in the transmitting assembly? (C&G)

8.4. (a) Draw circuit diagrams for each of the following devices used in line transmission systems: (i) balanced modulator, (ii) doubled-balanced modulator. (b) Give descriptive waveforms showing the principles of operation and explain the essential differences in the outputs obtained. (C&G)

FREQUENCY-DIVISION AND TIME-DIVISION MULTIPLEX SYSTEMS

8.5. (a) Explain, with the aid of a frequency spectrum diagram, how more than one signal can be transmitted over a common path using f.d.m. (b) Draw the spectrum diagrams of a CCITT 12-channel group.

8.6. With the aid of a block diagram explain the assembly of (a) a hypergroup and (b) a 2700-channel system. What is meant by the terms *erect* and *inverted* when applied to sidebands?

8.7. What is meant by the term *quantization* used in connection with pulse code modulation?

A 2000 Hz sine wave of peak value 7 V is to be transmitted by a p.c.m. system. The sampling frequency employed is 2.5 times the theoretical minimum value. (i) Determine the instantaneous value of the sine wave at each sampling instant. (ii) Assuming that 16 V is the maximum allowable peak–peak voltage permitted and that 16 sampling levels are used, draw the p.c.m. waveform which is transmitted to line. State any other assumptions made.

8.8. Explain how pulse code modulation differs from other types of pulse modulation and why it can be truly described as a digital system. Draw and explain the block diagram of a 4-channel p.c.m. t.d.m. system.

Short Exercises

8.9. (a) State the disadvantage of space division as the distance apart and number of users of a telephone network increase. (b) Compare the relative advantages and disadvantages of frequency- and time-division multiplex.

8.10. Using frequency spectrum diagrams explain how two similar baseband signals can be simultaneously transmitted over a common path using t.d.m.

8.11. Using frequency spectrum diagrams, explain how two similar baseband signals can be simultaneously transmitted over a common path using f.d.m.

8.12. Explain what is meant by each of the following terms: (i) supergroup, (ii) hypergroup, (iii) mastergroup.

Numerical Answers to Exercises

1.3 2.97 m   1.4 500 m, 33.3 m, 6 m   1.12 8.6 m
1.14 6×10^6 m, 3×10^3 m, 300 m, 3.33 m, 0.05 m   1.17 13.02 MHz
2.1 6800 Hz, 104 kHz, 100.6 kHz–103.7 kHz, 104.3 kHz–107.4 kHz
2.3 9 kHz, 501.5 kHz–505.95 kHz, 506.05 kHz–510.5 kHz
2.5 (a) 999 kHz, 1001 kHz, 0.25 V; (b) 999 kHz, 1001 kHz, 0.5 V
2.6 $m = 25\%$, V = 1.2 V, 310 kHz, 315 kHz, 305 kHz
2.10 (c) 1997–1999.95 kHz, 2000.05–2003 kHz, 6 kHz; (d) 100 m
2.11 $V_c = 4$ V, $V_m = 2$ V, 50%
2.13 $P_{LSF} = 100$ W, $m = 63.24\%$, 1000 W
2.14 25 kHz, 5 kHz   2.28 10 kHz, 16 kHz   3.8 1800 Ω, 34.72 μW
2.15 50 kHz   2.29 2 kHz   3.14 11.18:1, 9.13:1, 7.91:1
2.23 100 kHz, 10 rad, 5 rad   3.1 180 kHz, 130 kHz   5.1 15.85 μV, 501.2 μV
2.24 120 W   3.6 83 Hz, 165 Hz   5.4 17.3 dB
2.25 3.162 μV   3.7 1.291:1, 60 mW
5.6 +30 dBm, 14 dB, 16 dB, 15 dB, 19 dB, 14 dB, 18 dB, 50 W, 17 dB
5.7 6.3 μV, 200 μV, 66.6×10^{-8} W
5.8 11.76 dB, +26.53 dBm
5.9 100 W, 0.1 W, 2 mW, 4 mV, 6.1 mV, 41.65 μW
5.10 2 V, 6.67 mW
5.11 16 dB, 316 W, 1.58 W, 0.79 mW, 31.45 mW, 17 dB loss
5.12 0.22 mW, 23.3 mW, 3.522 W, 0.827 mW, 42 dB
5.15 −7.72 dBm
5.16 6.02 dB, 9.03 dB, 12.04 dB, 20 dB, 23 dB, 40 dB
5.17 12.04 dB, 18.06 dB, 24.08 dB, 40 dB, 46 dB, 80 dB
5.18 500 μW, 5 mW, 9.98 mW, 19.91 mW, 0.5 W
5.19 12.53 μW, 25.06 μW, 1 mW
5.20 +16.99 dBm, +23.01 dBm, +30 dBm, +36.99 dBm
5.21 0.1 mW, 0.398 mW, 1 mW, 10 mW, +10 dBm,
      +16 dBm, +20 dB, +30 dBm, 10 mW, 39.8 mW,
      100 mW, 1 W
5.22 25.12 μW, 5.02 μW, 1.26 μW
5.23 31.47 mW
5.24 2.194 dB/km
5.25 3.01 dB, 6.02 dB, 10 dB, 20 dB, 30 dB
6.3 60 dB
6.4 44.46 dB   6.14 106.56 dB
6.6 3.07 km   6.15 4.17 mA
6.8 4.82 MHz   6.16 5.755 km
6.9 7.47 dB   6.25 750 Ω
6.13 max. 44.77 km   6.27 257.14×10^3 km/sec

Index

Active filter 71
Amplified circuit
  four-wire 118
  two-wire 115
Amplitude modulation 22, 25
  bandwidth required for 27
  frequencies contained in 27
  merits of, relative to frequency modulation 40
  modulation factor or depth 30
  power content of 33
  principles of 25
  sidebands of 28
  sidefrequencies of 27
  single-sideband operation of 35
  use for data transmission 43

Balanced modulator 124
Bandpass filter 64
Bandstop filter 64
Bandwidth
  commercial speech circuits 11
  facsimile telegraphy 11, 51
  m.c.v.f. telegraphy 51
  music 12
  point-to-point radio links 11
  sound broadcasting 12
  telegraphy 10
  television broadcasting 8
  12 channel carrier telephony group 50
Bit 42

Cables
  construction of 98
    2.6/9.5 coaxial 101
    1.2/4.4 coaxial 102
    flexible coaxial 102
    internal 105
    star quad 98
    submarine 104
    unit twin 99
  equalization of 109
  loading of 106
Capacitance, of a line 89
Carrier frequency
  of radio systems
    high-frequency point-to-point 54
    mobile 55
    radio relay 56
    satellite links 57
    sound broadcasting 53
    television broadcasting 54
  of telegraphy systems 50
  of telephony systems 48
Characteristic impedance, of a line 90
Crystal filter 68

Data transmission 41
  use of amplitude modulation 43
  use of frequency shift 44
  use of phase modulation 45

Decibel 73
  definition of 75
  reference levels 79
  relationship with the neper 81
  use of voltmeter to measure 82
  voltage and current ratios 77
Deviation ratio, of a frequency-modulated wave 39

Equalizer 109

Facsimile (picture) telegraphy 10, 51
Feeders, radio 108
Filters 62
  active 71
  bandpass 64
  bandstop 64
  crystal 68
  high pass 64
  low pass 63
  m-derived 64
  modern designs of 68
Filters connected in parallel 70
Four-wire circuits 118
Frequencies
  commercial speech 11
  music 2
  speech 3
  television 7
Frequency bands, of radio waves 53
Frequency division multiplex 18, 124
  merits of, relative to t.d.m. 135
Frequency modulation 22, 37
  bandwidth required for 40
  deviation ratio of 39
  frequencies contained in 39
  frequency deviation 38
  merits of, relative to amplitude modulation 40
  modulation index 39
  principles of 37
  rated system deviation 38
Frequency shift modulation (data) 44

Group delay 94
Group velocity, of a line 94

Hearing 2
  frequency range of 3
High-pass filter 64
Hybrid coil 116
Hypergroup 130

Inductance, of a line 89
International telephone circuit 121

Leakance, of a line 89
Loading, of cables 106
Local line network 114
Low-pass filter 63

Mastergroup 131
Maximum power transfer 57

Modulation,
  amplitude 22, 25
  frequency 22, 37
  phase 22
  pulse 24
  pulse code 132
Modulators, balanced 124
Modulation factor 30
Modulation index 39
Morse code 8
Multi-channel telephony systems, 49
  hypergroup 130
  mastergroup 131
  supergroup 130
  supermastergroup 132
  twelve-channel group 127
Murray code 9
Music, frequencies contained in 2

Negative-impedance convertor 118
Neper, definition of 81
  relationship with the decibel 81

Pascal 3
Phase-change coefficient, of a line 93
Phase modulation 22, 44
Phase velocity, of a line 12, 94
Picture (facsimile) telegraphy 10, 51
Pulse modulation 24
Pulse code modulation 132

Rated system deviation, of a frequency-modulated wave 38
Resistance, of a line 89

Sideband, of amplitude-modulated wave 28
Sidefrequency, of amplitude-modulated wave 27
Signal-to-noise ratio 34
Single sideband operation, of amplitude-modulated wave 35
Speech, frequencies contained in 2
Supergroup 130
Supermastergroup 132

Telegraph codes 8
Telephone networks,
  local line 114
  trunk 113
Television, frequencies contained in 6
Terminating set 116
Time-division multiplex 20
  merits relative to frequency-division multiplex 135

Velocity of propagation on a line,
  group 94
  phase 12, 94

Wavelength, relationship with frequency 13

(A) Frequency Ranges

p. 48   (1) *Understands the terminology and states the frequencies used in line and radio telecommunications systems.*

11   1.1 Lists the frequency bands in common use for line and radio systems.
13   1.2 States the relationship between c, f and λ of an EM wave in free space.
13   1.3 Performs simple calculations using $c = f\lambda$.
13   1.4 States that the velocity of propagation along a line is less than c.
48   1.5 States the appropriate bandwidth of telegraph, speech, music and video baseband signals.
48, 128   1.6 States the relationship between baseband signal and carrier frequency.

(B) Multiplexing

18, 20   (2) *Describes how multiplexing is used to enable many users to transmit information simultaneously over a common path.*

124   2.1 States the disadvantage of space division as the distance and number of users increase.
18, 130   2.2 Draws frequency domain diagrams to explain how two similar baseband signals can be simultaneously transmitted over a common path using FDM.
20, 135   2.3 Draws time-domain diagrams to explain how two similar baseband signals can be simultaneously transmitted over a common path using TDM.
135   2.4 Compares the advantages and disadvantages of FDM and TDM.
130   2.5 Draws a frequency domain diagram of a CCITT 12-channel group.
130   2.6 States that a supergroup contains five groups.
130   2.7 States that a hypergroup contains a number of supergroups.
20, 132   2.8 Explains how an analogue signal can be sampled at regular intervals, and the information transmitted in digital form.

(C) Principles

73   (3) *Applies logarithmic units for ratios of power, voltage and current and appreciates the condition for maximum power transfer.*

73   3.1 Derives from "black-box" models, the power gain or attenuation of a telecommunications system.
75, 81   3.2 States the advantages of a system of logarithmically based units.
79   3.3 Defines the decibel and expresses levels in decibels relative to 1 mW.
76   3.4 Calculates the power gain/attenuation in cascaded systems.
57   3.5 States the maximum power transfer theorem (resistive conditions only).
77   3.6 Calculates the voltage or current gain/attenuation in dB.
34   3.7 Expresses signal-to-noise ratio in dB, given signal and noise power.

(D) Filters

62   (4) *Understands the effects and use of filters in telecommunication systems.*

62   4.1 States the function of a filter.
63, 65   4.2 Recognises, from a frequency plot, the type of filter being used (from low-pass, high-pass, band pass, band stop).
62   4.3 Draws the latest BS symbols for the various filter types and recognises superseded symbols.

(E) Modulation

18   (5) *Understands the concept of amplitude and frequency modulation in telecommunication systems.*

18   5.1 States the function of a carrier.
25   5.2 Explains, with the aid of a frequency spectrum graph, the production of a double side-band AM signal.
35   5.3 Explains, with the aid of a frequency spectrum graph, the production of a single-sideband by the elimination of the carrier and one sideband.
36   5.4 Compares double and single-sideband systems with specific reference to power and bandwidth.
33   5.5 Recognises the effect of over-modulation using time-domain waveforms.
35   5.6 Explains the relationship between bandwidth and signal-to-noise ratio.
37   5.7 Discusses the merits of frequency modulating a carrier wave.
37   5.8 Draws a representation of a frequency modulated wave.
   5.9 Explains the terms:
38   (*a*) frequency deviation
39   (*b*) frequency swing
39   (*c*) modulation index.
40   5.10 States that generally FM requires a greater bandwidth than AM.

41   (6) *Understands that amplitude and frequency modulation are used for data transmission.*

42   6.1 Explains by means of a given sketch why digital signals cannot be transmitted by communication systems.
43   6.2 Describes examples of modulation methods used for data transmission.

(F) Cables

98   (7) *Describes cables used in telecommunication systems.*
   7.1 Recognises the following types of cable from given cross-sectional diagrams:
105   internal multi-pair and multi-triple cables,
103   multi-tube coaxial cable,
98, 99   twin and quad type external cables.

109   (8) *Understands the effect of cables on analogue and digital signals.*

99, 102, 107   8.1 Sketches typical attenuation/frequency and group delay/frequency curves of unloaded, loaded and coaxial cables.
92, 94   8.2 Explains the shape of the curves in 8.1 in terms of the parameters of the cable.

(G) 2/4 Wire Working

115   (9) *Understands the concept of 2/4 wire working.*
118   9.1 Explains the need for separate paths for transmissions and reception.
115, 116   9.2 Draws a block diagram showing 2/4 wire conversion units with balance and line amplifiers.
116   9.3 States that an I/C signal divides equally between send and receive pairs at the transmit end and between the balance and the 2-wire at the receive end.
117, 118   9.4 Explains how a 4-wire circuit could become unstable.